Teaching and Learning Quantum Theory and Particle Physics

Teaching and Learning Quantum Theory and Particle Physics

Guest Editors

Oliver Passon
Gesche Pospiech
Thomas Zügge

MDPI

Basel • Beijing • Wuhan • Barcelona • Belgrade • Novi Sad • Cluj • Manchester

Guest Editors

Oliver Passon
School of Mathematics and Natural Sciences
Bergische Universität Wuppertal
Wuppertal
Germany

Gesche Pospiech
Faculty of Physics
TU Dresden
Dresden
Germany

Thomas Zügge
Institute of Physics
University Greifswald
Greifswald
Germany

Editorial Office
MDPI AG
Grosspeteranlage 5
4052 Basel, Switzerland

This is a reprint of the Special Issue, published open access by the journal *Physics* (ISSN 2624-8174), freely accessible at: https://www.mdpi.com/journal/physics/special_issues/TLQTPP.

For citation purposes, cite each article independently as indicated on the article page online and as indicated below:

Lastname, A.A.; Lastname, B.B. Article Title. *Journal Name* **Year**, *Volume Number*, Page Range.

ISBN 978-3-7258-5679-4 (Hbk)
ISBN 978-3-7258-5680-0 (PDF)
https://doi.org/10.3390/books978-3-7258-5680-0

Contents

About the Editors

Oliver Passon

Oliver Passon studied physics, mathematics, philosophy and education at the University of Wuppertal. After completing his PhD in experimental elementary particle physics at the European Laboratory for High Energy Physics (CERN) near Geneva, he was scientific staff at the Research Centre Jülich. In 2008 he completed his teacher training (mathematics and physics for secondary school) and worked for several years at a secondary school. Since 2013, he has been a member of the Physics Education group at the University of Wuppertal.

Gesche Pospiech

Gesche Pospiech is full professor of Physics Education at the Technische Universität Dresden. She earned her Ph D in Applied Mathematics at the University of Heidelberg and her habilitation in Physics Education at University of Frankfurt. In addition, she is fully trained teacher for mathematics and physics and has taught several years at Gymnasium. Her main research interest concerns modern physics in secondary school with a focus on quantum theory from an actual perspective. The second field of research is mathematics in physics education with emphasis von secondary school. In addition, she studies the pedagogical content knowledge of teachers in this field. She has been part in European projects fostering the science interest of students and in the Network Particle world in Germany. In addition, she is responsible for the physics education training of teacher students at TU Dresden.

Thomas Zügge

Thomas Zügge is responsible for the physics teacher education at the University of Greifswald, Germany. Before he completed his PhD in educational sciences/physics education in Wuppertal. As well experienced educator he facilitated workshops and trainings for a wide variety of organizations. His research interests include the education of modern physics, the application of developmental psychology in school and physics education research.

Preface

Quantum physics is a well-established part of many high-school curricula all over the world and the advance of quantum technologies (usually referred to as "second quantum revolution") may provide an additional boost. At the same time, even some elements of the standard model of particle physics are increasingly included in high school and undergrad syllabi.

Although the integration of these modern topics is welcome, it also poses a challenge for physics education. What is needed, initially, is a proper simplifications and educational reconstructions of the aforementioned topics. However, besides these basics the educational value and individual relevance ascribed by the students also need to be discussed and clarified.

At the high school and undergrad level, one is clearly not aiming at technical mastery of the formalism but more interested in conceptual clarity. This establishes a connection to issues in the philosophy and history of physics. Eventually, one should not lose sight of the established demand to integrate the "nature of science" (roughly speaking over and above the mere scientific facts) into the teaching of physics.

Oliver Passon, Gesche Pospiech, and Thomas Zügge
Guest Editors

physics

MDPI

Article

The Quasi-History of Early Quantum Theory

Oliver Passon

School of Mathematics and Natural Sciences, Bergische Universität Wuppertal, Gaußstr. 20, 42119 Wuppertal, Germany; passon@uni-wuppertal.de

Abstract: While physics has a rather ahistoric teaching tradition, it is common to include at least anecdotal reference to historical events and actors. These brief remarks on the history are typically distorted. I take issue with the textbook narrative of the historical development of early quantum theory and rectify some of the more severe misrepresentations. This seems to be all the more important, since the history of physics is commonly (and rightly) regarded as a means to foster scientific literacy and a more appropriate understanding of the nature of science (NoS).

Keywords: history of physics; Planck's law; light-quantum hypothesis; photo-effect; Bohr's atomic model; Compton effect; photon; nature of science (NoS)

1. Introduction

Although science teaching mainly aims at imparting current practices and theories, it is often framed historically. However, this "quasi-history" [1,2] often distorts the actual course of events and presents the history of science as a cumulative sequence which finally led to the acceptance of the current theories [3]. Thus, textbooks produce a sense of participation in a certain methodological and social tradition which may be fictitious [4]. In view of current debates in physics education, this may be characterized as a failure to provide an appropriate view on the nature of science (NoS).

This paper deals with the quasi-history of the discovery of energy quantization and the quantum of radiation, i.e., the "photon") as a case study (Section 2). As a by-product, my investigation helps to trace another problematic issue in the field which is disturbing beyond the question of historical accuracy. I mean the naïve line of tradition which is typically drawn from Einstein's tentative light-quantum hypothesis in 1905 to the current understanding of "photons" as objects described by quantum electrodynamics (QED). The decisive differences between the early light-quantum and the QED photon are often glossed over. As discussed in Section 2.5, this quasi-history of early quantum physics distorts both the history and the physics of the photon.

In closing (Section 3), I suggest a brief but more appropriate representation of the historical events and provide some context within the current NoS debate in physics education.

2. The Quasi-History of the Early Quantum Theory

This case study investigates the quasi-history of the discovery of energy quantization and the quantum of radiation (i.e., the "photon") in the early quantum theory as presented in physics textbooks. First, I sketch such a "typical" account which is a collage of elements frequently met (to substantiate my claim further, include references and quotations from selected textbooks; it should be emphasized that the discussion is not intended to denigrate these otherwise recommendable texts):

- Quantum mechanics arose in 1900 from the explanation of the black-body spectrum. To avoid the (ultra-violet) UV-catastrophe of the classical prediction, Max Planck introduced the quantum of energy, $\epsilon = h\nu$, where h is Planck's constant and ν denotes the frequency.

- In 1905 Albert Einstein generalized Planck's idea to the radiation field and introduced his light-quantum (with energy $E = h\nu$) to explain the photoelectric effect—a phenomenon which cannot be explained with the classical wave-theory of light.
- Niels Bohr in his atomic model from 1913 combined ideas from both Planck and Einstein. He proposed stable orbits for the electrons with fixed energies E_i. Any transition between these states is accompanied by an emitted (or absorbed) photon which meets the Bohr frequency condition $\Delta E = E_i - E_j = h\nu$.
- Arthur H. Compton's discovery of the effect named after him (i.e., the shift in wavelength of scattered X-rays) in 1922 and his explanation in terms of scattered light-quanta (soon to be called "photons") in 1923 convinced the last skeptics of Einstein's light-quantum hypothesis.

In showing how these steps oversimplify the actual course of events and contain even several plain errors, I can build on extensive scholarly work in the history of physics. Different aspects of this history have been closely investigated by [5–8] to mention just a few. More comprehensive overviews on the history of quantum theory have been given by, e.g., Jammer [9] and Mehra and Rechenberg [10].

Before starting, I would like to emphasize that while I try to correct several distortions and errors in the common quasi-history, I certainly do not intend to provide the "true" history of the events. The description surely fails to do full justice to the original motivation of the protagonists. The social dimension and will also oversimplify the development. In short, that this discussion also represents a specific narrative and some sort of "presentism", "Whiggism" [11], or "anachronism" [12] is both inevitable and maybe even desirable. However, the common quasi-history of quantum physics paints a naïve picture of a cumulative process in which anomalies ("UV-catastrophe"), successful predictions (photo-effect), or crucial experiments (Compton-scattering) drive the development. It is such a picture which is at odds with current views on the Nature of Science, i.e., my goal is to provide the necessary amendments to correct this *specific* misrepresentation.

Sections 2.1–2.4 deal with the abovementioned steps in the quasi-historical narrative.

2.1. Planck and the Black-Body Radiation Problem

Already in 1859, Gustav Kirchhoff argued that a black-body (i.e., a body that absorbs all incident radiation) in thermal equilibrium should emit a radiation spectrum which does not depend on its shape or material. Thus, there should be a universal function, $u = u(T, \nu)$, describing its spectral energy density at temperature T in the frequency interval $[\nu, \nu + d\nu]$—an ideal testing ground for theoretical models. In 1896, Willy Wien suggested a radiation law from thermodynamic arguments [13]:

$$u_{\text{Wien}} = \frac{8\pi\nu^2}{c^3} \cdot \frac{\alpha\nu}{\exp(\beta\nu/T)}, \tag{1}$$

where c denotes the speed of light. With the free parameters α and β adjusted accordingly, this expression could describe the available data well. Discrepancies showed up in the late summer of 1900.

Another specific theoretical prediction for the black-body spectrum is the so-called Rayleigh-Jeans (R–J) law, which follows from applying the equipartition theorem to the continuous Maxwell radiation field:

$$u_{\text{R–J}} = \frac{8\pi\nu^2}{c^3} \cdot kT, \tag{2}$$

where k is the Boltzmann constant.

However, the R–J law cannot be completely valid since it diverges in the high-frequency regime ("UV catastrophe"). Now, many textbooks (e.g., by Tipler and Llewellyn [14] (p. 122f)) suggest that the failure of this "classical" prediction was the great puzzle which initiated Planck's discovery of his radiation law in 1900 [15] (see [16] for English translation):

$$u_{\text{Planck}} = \frac{8\pi\nu^2}{c^3} \cdot \frac{h\nu}{\exp\left(\frac{h\nu}{kT}\right) - 1}, \tag{3}$$

including the discontinuous energy element, $\epsilon = h\nu$. This claim is incorrect for various reasons—the simplest being that the R–J law was only published in 1905, i.e., five years after Planck's law [17]. It is true, however, that Lord Rayleigh had anticipated this law already in 1900 in a brief note but introduced an exponential damping factor to avoid the divergence [18]. Planck did not cite Rayleigh's paper in his publications on the radiation problem from 1900 or 1901. However, given that Rayleigh's paper was cited by the experimenters (e.g., Rubens and Kurlbaum [19]), Planck must have known it. In any event, Planck's law displayed the $u \propto T$ relation for large wavelength which was found experimentally and also suggested by Rayleigh's argument.

Rayleigh returned to this issue only in 1905 and also provided the numerical factors which were missing in his publication from 1900. However, he committed a small mistake, and his result was too big by a factor of eight. James Jeans corrected this mistake immediately—thus the R–J law got its double name. The role of Jeans is discussed by McCaughan [20] who even suggested one possible origin of the quasi-historical narrative: It was Jeans himself who reported for the Physical Society of London on the history of the radiation problem in 1914. With respect to the R–J law, he stated: "This formula was given by Lord Rayleigh and the present author in 1900 [...]" [21]. This wrong date also entered the (much extended) second edition of the report in 1924 and made a textbook career thereafter. Another curious mistake can be found in the textbook of Alexander Komech. He claims without reference that the R–J law was already published in 1894 [22] (p. 12). Now, Jeans was a gifted student but entered Trinity College in Cambridge only at the age of 19 in October 1896. The common claim (see, e.g., [22], p. 15) that Planck in 1900 interpolated the radiation laws of Wien and R–J is obviously untenable for the same reason). To call the R–J law the "classical" prediction is also misleading, since the application of the equipartition theorem was debated at that time. For example, the failure of the rule of Dulong–Petit for the specific heat compromised this theorem, and it is well documented that it played no role in Planck's paper [5]; the term "catastrophe in the ultra-violet" was coined by Ehrenfest only in 1911.

In addition, it is unclear whether Planck in 1900 intended any *physical* energy quantization at all, a debate initiated by Thomas S. Kuhn in 1978 [6]. In Passon and Grebe-Ellis [23], a reconstruction of the different strands in this complex debate is provided.

Summing up, according to current historiography, Planck's study was no reaction to an anomaly, and it is doubtful whether Planck intended any quantization at all. However, be that as it may, nobody in the community picked up on "quantization" for many years anyway. Helge Kragh notes [24] (p. 63):

> "If a revolution occurred in physics in December 1900, nobody seemed to notice it, least of all Planck. During the first five years of the century, there was almost complete silence about the quantum hypothesis, which somewhat obscurely was involved in Planck's derivation of the black-body radiation law. The law itself, on the other hand, was quickly adopted because of its convincing agreement with experiment."

This is not the reaction one would expect if the very foundation of "classical physics" has just been shattered. Part of the explanation lies in Planck's rather obscure derivation of his law (it is evidently this obscurity, which fuels the controversial debate on its interpretation, I indicate above). In addition, one observes how claims of validity are negotiated within the scientific community. The problem of black-body radiation was a rather specific one, and the finer details of this study were of little concern to many. Kragh [24] (p. 69f) points out that Einstein's later application of Planck's distribution law to the problem of specific heat in 1907 (later extended by the Dutch theoretician Peter Debye [25]) played an important role since this was a more traditional field of physics. However, earlier than that Einstein introduced his famous light-quantum hypothesis; the second step in my reconstruction of the quasi-history.

2.2. Einstein and the Light-Quantum Hypothesis

In 1905, Einstein published several remarkable papers. Among them is the famous "Concerning an heuristic point of view toward the emission and transformation of light" [26] (see English translation in [27]). The usual textbook account caricatures its content as follows [28] (p. 12f):

> "The hypothesis of the existence of the quantum of light was introduced by Albert Einstein in 1905 starting from Planck's solution to the black-body problem. In this way, he was able to explain the photoelectric effect, i.e., the emission of electrons by a metal surface when it is illuminated by light."

or [14] (p. 129):

> "Einstein assumed that the energy quantization used by Planck in solving the black-body radiation problem was, in fact, a universal characteristic of light. Rather than being distributed evenly in the space through which it propagated, light energy consisted of discrete quanta, each of energy hf."

Historians of physics are quick to point out that almost every statement in these accounts is imprecise or wrong; Einstein's reference to Planck's radiation law is only marginal, the photoelectric effect is only one application mentioned (not even very prominently), and it is not "explained" but predicted, given that conclusive data became available only in 1914. Let us briefly set the record straight.

Einstein motivated his investigation with the "profound formal difference between the theoretical conceptions physicists have formed about [...] ponderable bodies, and Maxwell's theory of electromagnetic processes [...]". While material bodies are thought to consist of a "finite number of atoms and electrons" the electromagnetic field is described by continuous spatial functions [27] (p. 86). Einstein suggested that this asymmetry might produce problems in situations where material bodies and radiation interact, e.g., in the case of black-body radiation. With hindsight, it appears that Einstein anticipates a full-blown "quantization program" here in which fields and matter are treated on the same footing. While this assessment would mean to recast past developments in terms of current theories, Einstein's remarks are surely long-sighted. In any event, the paper is so good that its presentation needs no distortion.

The main body of the paper began by treating exactly this black-body radiation problem. Here, Einstein derived (independently) the "Rayleigh-Jeans law" (as it is now called) and emphasized the divergence of the total energy. Since this derivation was based on Maxwell's theory and (statistical) mechanics, it illustrates the problems caused by the asymmetry mentioned above. Einstein noted further that Planck's law approximates this expression in the large wavelength regime which allowed for the determination of the Avogadro constant. In the remainder of the paper, Einstein was silent on Planck's law.

The mathematical core which led to the introduction of light-quanta was based solely on Wien's radiation law (Equation (1)) from 1896. Einstein derived the entropy S_0 of the radiation in the frequency interval $[\nu, \nu + d\nu]$ contained in a volume V_0. A similar expression S holds for the entropy of a sub-volume V. From the entropy difference $S - S_0$, Einstein calculated the probability of an energy fluctuation into the sub-volume V:

$$W_{\text{rad}} = (V/V_0)^{\frac{N}{R}\frac{E}{\beta\nu}}. \tag{4}$$

Here N denotes the Avogadro constant and R the gas constant. Einstein then calculated the corresponding probability of n independent particles fluctuating into a sub-volume:

$$W_{\text{gas}} = (V/V_0)^n. \tag{5}$$

Comparing these equations, Einstein suggested that [27] (p. 97)

> "[...] monochromatic radiation of low density (within the range of validity of Wien's radiation formula) behaves thermodynamically as if it consisted of mutually independent energy quanta of magnitude $R\beta\nu/N$."

The wording was chosen carefully, i.e., there was no claim about, say, light in general consists of quanta, since the analysis dealt with a specific frequency regime only. However, Einstein introduced light-quanta heuristically as localized and distinguishable ("independent") entities.

Note also, that not even Planck's constant h was used in that paper. However, with $\beta = h/k$ and $k = R/N$, the exponent of Equation (4) turns into the familiar expression $E/h\nu$.

Einstein closed the paper with short sections on possible applications. These were (i) the Stokes rule of photo-luminescence, (ii) ionization of gases by ultra-violet light, and (iii) the photoelectric effect. Here the famous "Einstein equation" (in modern notation) $T_{\text{max}} = h\nu - W$ was introduced, with T_{max} the kinetic energy of the photo-electrons and W the work function (I note in passing,, that the actual measurements (e.g., Millikan's stopping potential method) are not sensitive to the work function of the cathode, i.e., the material that emits the electrons. Due to contact potentials, one surprisingly measures rather the work function of the anode [29]).

While Millikan could confirm this Einstein equation in 1914 [30], this result did not even convince himself of the validity of the light-quantum hypothesis. Apparently most physicists joined him in this assessment because there were also competing models to account for the data [7]. In addition, Millikan's experiment measured only the maximal energy of the photo-electrons and not the energy spectra. It was the paper by Maurice de Broglie in 1921 which removed this ambiguity and had central importance for the growing acceptance of the light-quantum hypothesis [31] (p. 266ff).

Summing up, the light-quantum hypothesis was no application or generalisation of Planck's paper on black-body radiation . The explanation/prediction of the photo-effect was no main intention of Einstein and the confirmation of the "Einstein equation" only played a minor role in the gradual acceptance of the light-quantum hypothesis. Not least important, the frequent claim that this effect has no explanation in terms of the wave-theory of radiation is likewise incorrect. In fact, the photoelectric effect can be described by the semi-classical approximation in which the radiation is treated classically (see Section 2.5). Importantly, this semi-classical explanation also gets the angular distribution of the photo-electrons right, while the naïve corpuscular explanation fails in this respect [32].

2.3. Bohr and the Atomic Model

Bohr's atomic model of 1913 is discussed by many textbook authors as an *application* of Einstein's light-quantum hypothesis. Douglas Giancoli provides a typical example [33] (p. 789):

> "In this Bohr model, light is emitted only when an electron jumps from a higher (upper) stationary state to another of lower energy. When such a transition occurs, a single photon of light is emitted."

A similar statement can be found, e.g., in [34], p. 7, or in [35], p. 76.

It is certainly true that Bohr's model puts electrons on stationary states with discrete energies E_i and that the transition between these states is accompanied by "homogeneous" (i.e., monochromatic) radiation of the frequency $\nu = \Delta E/h$ [36]. However, Bohr took this to be just classical electromagnetic radiation which happens to have a specific frequency, and his rejection of the light-quantum hypothesis until the mid 1920s is well documented [10] (pp. 532–554). In his trilogy from 1913, Bohr stated explicitly that he tried to offer a solution which remained consistent "with experiments on phenomena for which a satisfactory explanation has been given by the classical dynamics and the wave theory of light" [36] (p. 19). This misrepresentation of Bohr's study was also noted by Stachel [37].

To view the Bohr model as a successful application of the light-quantum hypothesis should confuse textbook authors and readers anyway. It is well known (and often remarked) that the light-quantum hypothesis was mostly rejected (until, say, the early 1920s), while the Bohr model was well received from the outset, could celebrate early successes, but passed its heydays in the early 1920s [38]. This would be rather odd (or even self-contradictory) if the latter would contain the former.

There is certainly a sense in which Bohr's model provided a radical departure from conventional electrodynamics since the radiation frequency is *different* from the orbital motion frequency. (Interestingly, Stachel [39] collects some evidence that Einstein may have, already anticipated Bohr's atomic model around 1905. In any event, he was very excited about its discovery. His derivation of Planck's radiation law from the study of transition processes [40] (see [41] for English translation) applied Bohr's idea of stationary states and provided another justification for the frequency condition $\Delta E = h\nu$—this time even for a non-periodic process.) However, as is well-known, the later developments (starting with Heisenberg's matrix mechanics) went in another direction and abandoned particle orbits and semiliteral models of the Bohr-type altogether. (As noted by Heilbron [42], the Bohr model was actually a late product of Victorian physics and stood in a line of work initiated by J. J. Thompson. Some of its success can be traced to the impact of the first world war. This great divide in modern Western history also interrupted research lines and precluded the development of alternatives. Note that much of the further developments of Bohr's model were conducted by Arnold Sommerfeld who escaped military service owing to his age. Heilbron [42] (p. 230) suggests that not only for general history is the first world war the watershed between the 19th and 20th centuries—but also for the history of physics).

Nicolaas P. Landsman remarks with respect to Bohr that "[h]is model probably would have gained in consistency by adopting the photon picture of radiation" [43] (p. 424). Thus, this Bohr episode provides a good example for the quasi-historical tendency to increase the coherence and to distort past events to fit them into a more rational narrative.

However, there is an additional twist to the Bohr story. When asked to contribute an article to the Einstein volume of the series *Living Philosophers*, Bohr famously wrote on his discussions with Einstein on epistemological problems. To set the stage, he briefly summarized the development of quantum theory including his atomic model from 1913. There one reads the following account of the radiation between stationary states [44] (p. 204):

> "[...] the spectra were emitted by a step-like process in which each transition is accompanied by the emission of a monochromatic light-quantum of an energy just equal to that of an Einstein photon."

This shortened account is clearly justified, given that Bohr aims at a different issue in this essay. While he speaks of a "light-quantum" he still does not call it "Einstein's light-quantum" but just quanta of the same energy as the "Einstein photon" (actually, Einstein did not use the term "photon" in any of his papers). Still, when considered superficially, this Bohr quotation apparently supports the quasi-historical account of the Bohr model. This nicely illustrates that quasi-historical accounts are also produced by the physicists involved in the development.

2.4. Compton and The Light-Quantum

Let us turn to the Compton effect, which allegedly played a central role in the final acceptance of the light-quantum hypothesis. For example, Tipler and Llewellyn [14] (p. 561) state:

> "Einstein's suggestion was not widely accepted until, over the next 20 years, Millikan's thorough experimental investigation of the photoelectric effect and Compton's discovery and explanation of the Compton effect provided incontrovertible evidence for the quantization of electromagnetic radiation, the field quantum being a particle we now call the photon."

In a similar vein is the following "historical remark" [28] (p. 35):

> "We would like to end this Subsection with a historical remark. Einstein's interpretation of the photoelectric effect involving the corpuscular nature of light had not completely convinced the scientific community about the quantization of the electromagnetic field. In this respect, the Compton effect, where energy and momentum are conserved in each single collision, played the role of a definitive experimental evidence of radiation quantization and convinced even the most skeptical physicists."

Brush [45] has pointed out that it is difficult to provide clear evidence for such acceptance claims. The problem is that in scientific publications the authors rarely explain if and (especially) why they have changed their minds on certain issues. Apparently, such a claim is often based on anecdotal evidence by a few prominent physicists. Brush [45] (Ch. 7.9) suggests another way to study the impact of the Compton effect on the acceptance of the light-quantum hypothesis. It is analysed in terms of the reception it found in textbooks and reviews of that period, i.e., texts with a stronger commitment to give an extended description. It can be shown that starting arround 1926, reviews quoted the Compton effect more often as compelling evidence for light-quanta, while textbooks and popular articles of that time favored the photoelectric effect.

However, the above quotations call Millikan's and Compton's study "incontrovertible" or "definitive experimental evidence" for the "quantization of electromagnetic radiation, the field quantum being a particle we now call the photon". Evidently these authors insinuate that the current reader should be equally convinced and that this is not just a matter of historical contingency but scientific necessity. However, the situation is more complex here, and apparently the view on the relation between Compton effect and the light-quantum hypothesis changed again after 1925/1926.

With the advent of matrix and wave-mechanics, the so-called semi-classical approximation could be applied to known phenomena (according to Alexander Blum [46], this was not viewed as an approximation at first). Here, the continuous (i.e., unquantized) Maxwell-field is inserted into the Hamilton operator, i.e., matter is treated quantum mechanically and the radiation field classically. Since this method allows a description of the photoelectric effect and Compton scattering [47,48], there is apparently no need to introduce any field quantization at this point.

The power of the semi-classical treatment extends even to more than the simple kinematics of the Compton scattering. Compton's original analysis left the *intensity distributions* (in energy and angle) open. This gap was filled by the study by Oskar Klein and Yoshio Nishina in 1929 [49]. However, their derivation (taking full account of relativity and spin) still applied the semi-classical approximation for the radiation, even though at the time the beginnings of QED were already developing [50] (p. 233f). Against this background, the following textbook account is particularly disturbing [14] (p. 141, Note 16):

> "It was Compton who suggested the name photon for the light-quantum. His discovery and explanation of the Compton effect earned him a share of the Nobel Prize in Physics in 1927."

Now, the invention of the name "photon" is usually attributed to the American chemist Gilbert N. Lewis [51]. Kragh [52] has discovered that this name was used before (and independently) by at least four different researchers. Priority belongs here to the American physicist and psychologist Leonard T. Troland, who already coined the expression in 1916 in his paper [53], published a year later. Note that to all of these scientists, "photon" did mean something different than Einstein's "quantum of light". Compton rightly deserves credit for the popularization of the term in its current meaning, and that was not a little thing: To call Einstein's light-quantum a "phot-on" on a par with "electr-on" or "prot-on" reflects the widespread recognition of this concept and, so to say, elevated it into into a higher rank. At the same time, the name "photon" reflects the tendency of reification of the underlying concept; cf. Section 2.5. However, more important is the second claim of the above quotation. In fact, it is not correct that Arthur Compton was awarded the Nobel prize for "discovery *and explanation*" of the Compton effect, but only for the "*discovery* of the effect named after him" as Karl Siegbahn stated in his presentation speech at the award ceremony [54]. Siegbahn's speech contains another interesting remark which reveals why the distinction between "discovery" and "explanation" really matters:

> "[...] the Compton effect has, through the latest evolutions of the atomic theory, got rid of the original explanation based upon a corpuscular theory. The new wave mechanics, in fact, lead as a logical consequence to the mathematical basis of Compton's theory. Thus the effect has gained an acceptable connection with other observations in the sphere of radiation." [54]

Karl Siegbahn delivered this speech as a member of the Nobel committee. However—more relevant in this context—he was an eminent physicist himself who had received the 1924 Nobel prize in physics for his discoveries and research in the field of X-ray spectroscopy.

The above quote, however, means that the corpuscular underpinning of the effect was questioned—with reference to the newly discovered wave mechanics, i.e., Schrödinger's theory. Compton's not receiving the prize for the explanation of his effect was surely no oversight but pure intention (note that this is as with Einstein, who received the Nobel prize in 1922 for his discovery of the law of the photoelectric effect—and not for the proposed explanation, i.e., the light-quantum hypothesis).

Siegbahn's remark appears to be very much in the spirit of the times, as noted by Brown [50] (p. 228). Again, such a reception claim is hard to verify from research papers alone. An interesting source is provided by a paper written by Otto Halpern and Hans Thirring in 1929 [55]. This paper is contained in the publication series *Ergebnisse der Exakten Naturwissenschaften* (*Results of the Exact Sciences*), which was intended to provide a semi-technical overview of current developments for students and physicists, working in neighboring areas. Given that this two-part paper has roughly 200 pages, it could be almost called a textbook. After discussing the semi-classical treatment of the photoelectric and Compton effects according to Beck [47] and Schrödinger [48], Halpern and Thirring remark [55] (p. 444):

> "According to current understanding, there is no compelling reason to accept the existence of "light atoms". In how far their acceptance could be based on general considerations concerning the wave-particle parallelism shall not be treated here. This question reaches into the area of quantum electrodynamic, a subject matter only now in the process of arising." (translated by myself)

Hence, the confusion about the light-quantum has recurred in 1926/27 after its apparently being already settled in 1923 (surprisingly, this point is merely mentioned incidentally in the standard work on the Compton effect by Stuewer [56] (pp. 288–290)). The important point here is the following: the photon concept which eventually "reappeared" through the development of quantum electrodynamic has distinctly other features than Einstein's early conception or Compton's photon in his kinematic treatment of the photon-electron scattering. It is certainly no accident that it got the same name, but for conceptual reasons, this appears unfortunate nevertheless. This development is discussed in Section 2.5 just below.

2.5. The Current Photon Concept

The above Subsections have concentrated on the misrepresentations of the history of the early energy- and light-quantum. However, there is another issue involved which is disturbing beyond the question of historical (in-)accuracy. It is the naïve line of tradition which is typically drawn from Einstein's tentative light-quantum hypothesis in 1905 to the current understanding of "photons" as objects described by QED. In fact, many authors downright identify Einstein's light-quantum with the current photon. For example, Giancoli states with reference to Einstein's 1905 paper that "[...] this idea suggests that light is transmitted as tiny particles, or photons as they are now called" [33] (p. 775). In a similar vein Tipler and Llewellyn [14] (p. 561) remark "the Compton effect provided incontrovertible evidence for the quantization of electromagnetic radiation, the field quantum being a particle we now call the photon". Auletta et al. [28] (p. 12f) claim that the "delocalized" classical wave is unable to account for the photoelectric effect. Emphasizing the "delocalized wave" is apparently suggesting that only a localized photon can account e.g., for the photoelectric effect.

However, as mentioned above, the early indications of light-quanta turned out to be explicable in terms of the semi-classical approximation, i.e., did not imply any quantization of the radiation field. As noted recently by Blum and Jähnert, it is curious that the very history of quantum physics was inspired by problems of radiation theory (black-body radiation, photo electricity, atomic spectra, dispersion etc.), however, the emerging theories of matrix- and wave mechanics turned out to be no theories of light and radiation—but theories of matter [57]. Within non-relativistic quantum mechanics, the photon is a foreign body anyway, and its technical foundation lies in quantum electrodynamics.

This is not the place to get into the specifics of quantum electrodynamics, but suffice it to note that there is no position operator for the photon [58] and that there is no "wavefunction" of the photon with probability interpretation in three-space [59]. The current photon cannot be localized—not even fuzzy. In addition, it is indistinguishable, while Einstein's paper from 1905 applied the statistics of distinguishable objects.

To call the photon a "particle" is rather a jargon and refers to the discreteness of the spectrum of the occupation number operator. It is rather an abstract notion which defies naïve reification. Genuine QED effects which need the field quantization (i.e., photons) for their explanation are subtle and well beyond the range of any ordinary quantum physics curriculum on high school or college level (e.g., spontaneous emission or the hyperfine splitting in hydrogen). Consequently, one is committing a technical (and not just historical) mistake if one suggests that the photoelectric effect or Compton scattering imply the need for the quantization of the electromagnetic field and that Einstein's localized and distinguishable light-quantum corresponds to what is now called photon.

This problem in the teaching of the photon concept has been long noted [60–62]. Apparently, it still prevails, and one reason for its persistence is surely the quasi-history dealt with above. There is obviously a great desire for physicists to endow their concepts with an honorable family-tree. Another famous example is provided by the concept of "atom" and "atomism". Many physicists trace the origin of modern atomism to the alleged origin in the 5th century BC, i.e., Leucippus and Democritus. However, the atoms of ancient Greece were certainly quite different from contemporary matter constituents and such a line of tradition is rather misleading.

3. Summary and Conclusions

At the beginning of Section 2, I sketch the common quasi-history of the early light-quantum with a four-item list of typical steps (i.e., the standard presentation of Planck's law, Einstein's light-quantum, Bohr's atomic model, and the Compton effect). Against the backdrop of my discussion, a more appropriate version of this list may sounds like this:

- In 1900, Planck introduced the energy element, $\epsilon = h\nu$, to account for the black-body spectrum. To avoid the so-called ultra-violet-he was not his intention. If Planck even intended a physical quantization is debated among historians of physics.
- Einstein's light-quantum hypothesis was not based on Planck's law but on Wien's law. His light-quanta should not be confused with the current photon concept since they were localized and distinguishable. The photo-effect did not play a prominent role in this paper either.
- Bohr's atomic model applies ideas of Einstein's theory of specific heat, but the light-quantum was rejected by Bohr until 1925. In Bohr's model, the radiation follows the frequency condition but is treated classically.
- The Compton effect (and its explanation in 1923) convinced many physicists of the reality of light quanta. With the advent of quantum mechanics in 1925/1926, the picture become more differentiated. Here, Compton effect and and photo-effect can be explained with the classical radiation field. A genuine quantum electrodynamics

(QED) effect which could be used to motivate the current photon would be, e.g., spontaneous emission.

Again, this brief account is not meant to represent the "true" history of the events. Textbooks need to simplify their subject matter, and they inevitably include inaccuracies or even errors. Additionally, in physics teaching, historical accuracy is no end in itself. In the same context, the historian of physics Helge Kragh rightly remarked [8] (p. 360):

> "The purist attitude that only detailed, scholarly acceptable history should enter textbooks, amounts in practice to a denial of a historical perspective in science teaching."

However, the common quasi-history of the light-quantum in early quantum theory contains not just accidental mistakes. It systemically paints a picture of physics research as a cumulative process in which anomalies ("ultra-violet catastrophe"), successful predictions (photo-effect) or crucial experiments (Compton-scattering) drive the development inevitably towards our current understanding. In short, it is a typical Whig history. In view of current debates in physics education this may be characterized as a failure to provide an adequate view of the nature of sciense (NoS).

NoS typically refers to knowledge "about" science, as opposed to mere scientific content knowledge. Thus, NoS includes epistemological, historical and sociological aspects of science and its making. Notwithstanding the universal endorsement of NoS within the physics education literature and curriculum reform documents the very meaning of NoS remains elusive and debated. A pragmatic solution to this problem was provided by Norman Lederman and his group [63,64] who championed the so-called consensus-based NoS-view. Several authors have criticized this view and emphasized the more controversial and context dependent character of NoS; see, e.g., [65–67]. However, all approaches of NoS agree that including elements from the history and philosophy of science (HPS) can promote the teaching of NoS.

Henke and Höttecke [68] discuss major obstacles for the successful implementation of HPS elements into the classroom. One of them is a lack of adequate HPS content (e.g., case-studies or vignettes) in textbooks. Now, this case study is a reminder that there is not only a lack of adequate content but a huge problem with inadequate material. I think that the current NoS debate should pay closer attention to this issue since this problem is not confined to quantum theory.

Funding: This research received no external funding.

Acknowledgments: I thank the anonymous referees and the editorial committee for their helpful critical comments.

Conflicts of Interest: The author declares no conflict of interest.

References

1. Whitaker, M.A.B. History and quasi-history in physics education. I. *Phys. Educ.* **1979**, *14*, 108–112. [CrossRef]
2. Whitaker, M.A.B. History and quasi-history in physics education—Part 2. *Phys. Educ.* **1979**, *14*, 239–242. [CrossRef]
3. Klein, M.J. The use and abuse of historical teaching in physics. In *History in the Teaching of Physics*; Brush, S.G., King, A.L., Eds.; University Press of New England: Hanover, NH, USA, 1972; pp. 12–18.
4. Brush, S.G. Should the history of science be rated X. *Science* **1974**, *183*, 1164–1172. [CrossRef] [PubMed]
5. Klein, M. Max Planck and the beginnings of quantum theory. *Arch. Hist. Exact Sci.* **1962**, *1*, 459–479. [CrossRef]
6. Kuhn, T.S. *Black-Body Theory and the Quantum Discontinuity, 1894–1912*; University of Chicago Press: Chicago, IL, USA, 1978. Available online: https://vdoc.pub/documents/black-body-theory-and-the-quantum-discontinuity-1894-1912-3mfkvonmfvd0 (accessed on 5 July 2022).
7. Stuewer, R.H. Non-Einsteinian interpretations of the photoelectric effect. In *Historical and Philosophical Perspectives of Science*; Stuewer, R.H., Ed.; University of Minnesota Press: Minneapolis, MN, USA, 1970; pp. 246–263. Available online: https://conservancy.umn.edu/bitstream/handle/11299/184655/5-11_Stuewer.pdf (accessed on 5 July 2022).
8. Kragh, H. A Sense of history: History of science and the teaching of introductory quantum theory. *Sci. Educ.* **1992**, *1*, 349–363. [CrossRef]

9. Jammer, M. *The Conceptual Development of Quantum Mechanics*; McGraw-Hill: New York, NY, USA, 1966. Available online: https://archive.org/details/conceptualdevelo0000jamm/page/n7/mode/2up (accessed on 5 July 2022).
10. Mehra, J.; Rechenherg, H. *The Historical Development of Quantum Theory*; Springer: New York, NY, USA, 1982; Volumes 1–4.
11. Hall, A.R. On Whiggism. *Hist. Sci.* **1983**, *21*, 45–59. [CrossRef] [PubMed]
12. Brush, S. Anachronism and the history of science: Copernicus as an airplane passenger. In *Scientia Poetica, Volume 8*; Danneberg, L., Kablitz, A., Schmidt-Biggemann, W., Thomé, H., Vollhardt, F., Eds.; Walter de Gruyter GmbH & Co. KG: Berlin, Germany, 2004; pp. 255–264. [CrossRef]
13. Wien, W. Ueber die Energievertheilung im Emissionsspectrum eines schwarzen Körpers. *Ann. der Phys.* **1896**, *294*, 662–669. [CrossRef]
14. Tipler, P.A.; Llewellyn, R.A. *Modern Physics*; W. H. Freeman and Company: New York, NY, USA, 2009. Available online: http://web.pdx.edu/~pmoeck/books/Tipler_Llewellyn.pdf (accessed on 5 July 2022).
15. Planck, M. Zur Theorie des Gesetzes der Energieverteilung im Normalspectrum. *Verh. Dtsch. Phys. Ges. Berlin* **1900**, *2*, 237–245. [CrossRef]
16. Planck, M. On the theory of the energy distribution law of the normal spectrum. In *The Old Quantum Theory*; Ter-Haar, D., Ed.; Pergamon Press Ltd.: Oxford, UK, 1967; pp. 82–91. Available online: https://webéducation.com/the-old-quantum-theory-pdf/ (accessed on 5 July 2022).
17. Jeans, J.H. On the partition of energy between matter and Æther. *Lond. Edinb. Dublin Philos. Mag. J. Sci.* **1905**, *10*, 91–98. [CrossRef]
18. Lord Rayleigh. Remarks upon the law of complete radiation. *Lond. Edinb. Dublin Philos. Mag. J. Sci.* **1900**, *49*, 539–540. [CrossRef]
19. Rubens, H.; Kurlbaum, F. Anwendung der Methode der Reststrahlen zur Prüfung des Strahlungsgesetzes. *Ann. der Phys.* **1901**, *309*, 649–666. [CrossRef]
20. McCaughan, J.B.T. Jeans' role in the law of black-body radiation. *Phys. Educ.* **1980**, *15*, 255–260. [CrossRef]
21. Jeans, J. *Report on Radiation and the Quantum–Theory*; "The Electrician" Printing & Publishing Company, Ltd.: London, UK, 1914. Available online: https://archive.org/details/cu31924012330407/page/n3/mode/2up (accessed on 5 July 2022)
22. Komech, A. *Quantum Mechanics: Genesis and Achievements*; Springer Science+Business Media: Dordrecht, The Netherlands, 2013. [CrossRef]
23. Passon, O.; Grebe-Ellis, J. Planck's radiation law, the light-quantum, and the prehistory of indistinguishability in the teaching of quantum mechanics. *Eur. J. Phys.* **2017**, *38*, 035404. [CrossRef]
24. Kragh, H. *Quantum Generations: A History of Physics in the Twentieth Century*; Princeton University Press: Princeton, NJ, USA, 1999. [CrossRef]
25. Debye, P. Zur Theorie der spezifischen Wärmen. *Ann. der Phys.* **1912**, *344*, 789–839. [CrossRef]
26. Einstein, A. Über einen die Erzeugung und Verwandlung des Lichtes betreffenden heuristischen Gesichtspunkt. *Ann. der Phys.* **1905**, *17*, 132–148. [CrossRef]
27. Einstein, A. On a heuristic point of view concerning the production and transformation of light. In *The Collected Papers of Albert Einstein. Volume 2: The Swiss Years: Writings, 1900–1909 (English translation supplement)*; Stachel, J.; Cassidy, D.C.; Renn, J.; Schulmann, R., Eds. Princeton University Press: Princeton, USA, 1989; pp. 86–103. Available online: http://einsteinpapers.press.princeton.edu/vol2-trans/100 (accessed on 5 July 2022).
28. Auletta, G.; Fortunato, M.; Parisi, G. *Quantum Mechanics*; Cambridge University Press: Cambridge, UK, 2009. [CrossRef]
29. Rudnick, J.; Tannhauser, D.S. Concerning a widespread error in the description of the photoelectric effect. *Am. J. Phys.* **1976**, *44*, 796–798. [CrossRef]
30. Millikan, R.A. A Direct determination of "*h*". *Phys. Rev.* **1914**, *4*, 73–75. [CrossRef]
31. Wheaton, B.R. *The Tiger and the Shark: Empirical Roots of Wave–Particle Dualism*; Cambridge University Press: Cambridge, UK, 1983.
32. Greenstein, G.; Zajonc, A.J. *The Quantum Challenge*; Jones and Bartlett Publishers, Inc.: Sudbury, MA, USA, 1997. Available online: https://webéducation.com/the-quantum-challenge-pdf/ (accessed on 5 July 2022).
33. Giancoli, D.C. *Physics: Principles with Applications*; Pearson Education, Inc.: Boston, MA, USA, 2014. Available online: https://www.docdroid.net/OFMOth4/giancoli-physics-principles-7th-ed-pdf (accessed on 5 July 2022).
34. Weinberg, S. *Lectures on Quantum Mechanics*; Cambridge University Press: Cambridge, UK, 2015. [CrossRef]
35. Griffiths, D.J. *Revolutions in Twentieth-Century Physics*; Cambridge University Press: Cambridge, UK, 2012.
36. Bohr, N.I. On the constitution of atoms and molecules. *Lond. Edinb. Dublin Philos. Mag. J. Sci.* **1913**, *26*, 1–25. [CrossRef]
37. Stachel, J. Bohr and the photon. In *Quantum Reality, Relativistic Causality, and Closing the Epistemic Circle*; Myrvold, W.C., Christian, J., Eds.; Springer Science+Business Media B.V.: Dordrecht, The Netherlands, 2009; pp. 69–83. [CrossRef]
38. Kragh, H. The many faces of the Bohr atom. *arXiv* **2013**, arXiv:1309.4200. [CrossRef]
39. Stachel, J. Einstein and the quantum: Fifty years of struggle. In *From Quarks to Quasars: Philosophical Problems of Modern Physics*; Colodny, R.G., Ed.; University of Pittsburgh Press: Pittsburg, PA, USA, 1986; pp. 349–381.
40. Einstein, A. Zur Quantentheorie der Strahlung. *Mitteil. Phys. Gesel. Zürich* **1916**, *18*, 47–62. [CrossRef]
41. Einstein, A. On the quantum theory of radiation. In *The Collected Papers of Albert Einstein. Volume 6: The Berlin Years: Writings, 1914-1917 (English Translation Supplement)*; Klein, M.J., Kox, A.J., Schulman, R., Eds.; Princeton University Press: Princeton, NJ, USA, 1997; pp. 220–233. Available online: http://einsteinpapers.press.princeton.edu/vol6-trans/232 (accessed on 5 July 2022).
42. Heilbron, J.L. Rutherford–Bohr atom. *Am. J. Phys.* **1981**, *49*, 223–231. [CrossRef]

43. Landsman, N.P. Between classical and quantum. In *Philosophy of Physics. Part A*; Butterfield, J., Earman, J., Eds.; Elsevier: Amsterdam, The Netherlands, 2007; pp. 417–553. [CrossRef]
44. Bohr, N. Discussions with Einstein on epistemological problems in atomic physics. In *The Library of Living Philosophers, Volume 7. Albert Einstein: Philosopher-Scientist*; Schilpp, P.A., Ed.; Open Court: Chicago, IL, USA, 1949; pp. 199–241. Reprinted in *Nils Bohr Collected Works*; Kalckr, J., Ed.; Elsevier Science B.V.: Amsterdam, The Netherlands, 1996; Volume 7; pp. 339–381. [CrossRef]
45. Brush, S.G. *Making 20th Century Science: How Theories Became Knowledge*; Oxford University Press: New York, NY, USA, 2015.
46. Blum, A. Private communication.
47. Beck, G. Comptoneffekt und Quantenmechanik. *Z. Phys.* **1926**, *38*, 144–148. [CrossRef]
48. Schrödinger, E. Über den Comptoneffekt. *Ann. der Phys.* **1927**, *387*, 257–264. [CrossRef]
49. Klein, O.; Nishina, Y. Über die Streuung von Strahlung durch freie Elektronen nach der neuen relativistischen Quantendynamik von Dirac. *Z. Phys.* **1929**, *52*, 853–868. [CrossRef]
50. Brown, L.M. The Compton effect as on path to QED. *Stud. Hist. Phil. Mod. Phys.* **2002**, *33*, 211–249. [CrossRef]
51. Lewis, G.N. The Conservation of photons. *Nature* **1926**, 118, 874–875. [CrossRef]
52. Kragh, H. Photon: New light on an old name. *arXiv* **2014**, arXiv:1401.0293. [CrossRef]
53. Troland, L. T. On the measurement of visual stimulation intensities. *J. Exp. Psych.* **1917**, *2*, 1–33. [CrossRef]
54. Nobel Prize in Physics 1927—Presentation Speech. Nobelprize.org. Nobel Media AB 2014. Web. 25 Aug 2016. 1927. Available online: http://www.nobelprize.org/nobel_prizes/physics/laureates/1927/press.html (accessed on 5 July 2022).
55. Halpern, O.; Thirring, H. Die Grundgedanken der neueren Quantentheorie. II. Teil. Die Weiterentwicklung seit 1926. In *Herausgegeben bei der Schriftleitung der "Naturwissenschaften" (Springer Tracts in Modern Physics), Volume 8*; Springer: Berlin/Heidelberg, Germany, 1929; pp. 367–508. [CrossRef]
56. Stuewer, R.H. *The Compton Effect: Turning Point in Physics*; Science History Publications: New York, NY, USA, 1975. Available online: https://archive.org/details/comptoneffecttur0000stue/page/n3/mode/2up (accessed on 5 July 2022).
57. Blum, A.S.; Jähnert, M. The birth of quantum mechanics from the spirit of radiation theory. *Stud Hist Philos Sci.* **2022**, *91*, 125–147. [CrossRef] [PubMed]
58. Newton, T.D.; Wigner, E.P. Localized States for Elementary Systems. *Rev. Mod. Phys.* **1949**, *21*, 400–406. [CrossRef]
59. Peierls, R. *Surprises in Theoretical Physics*; Princeton University Press: Princeton, NJ, USA, 1979.
60. Simonsohn, G. Probleme mit dem Photon im Physikunterricht. *Prax. Naturwissenschaften* **1981**, *30*, 257–266.
61. Strnad, J. Photons in introductory quantum physics. *Am. J. Phys.* **1986**, *54*, 650–652. [CrossRef]
62. Kidd, R.; Ardini, J.; Anton, A. Evolution of the modern photon. *Am. J. Phys.* **1989**, *57*, 27–35. [CrossRef]
63. McComas, W.F.; Clough, M.P. Nature of science in science instruction: Meaning, advocacy, rationales, and recommendations. In *Nature of Science in Science Instruction. Science: Philosophy, History and Education*; McComas W., Ed.; Springer: Cham, Switzerland, 2020; pp. 3–22. [CrossRef]
64. Lederman, N.G. Nature of science: Past, present, and future. In *Handbook of Research on Science Education*; Abell, S.K., Lederman, N.G., Eds.; Lawrence Erlbaum Associates: Hillsdale, MI, USA, 2007; pp. 831–880. [CrossRef]
65. Duschl, R.A.; Grandy, R. Two Views About Explicitly Teaching Nature of Science. *Sci. Educ.* **2013**, *22*, 2109–2139. [CrossRef]
66. Matthews M.R. Changing the focus: From nature of science (NOS) to features of science (FOS). In *Advances in Nature of Science Research*; Khine M.S., Ed.; Springer: Dordrecht, The Netherlands, 2012; pp. 3–26. [CrossRef]
67. Allchin, D. Evaluating knowledge of the nature of (whole) science. *Sci. Educ.* **2011**, *95*, 518–542. [CrossRef]
68. Henke, A.; Höttecke, D. Physics teachers' challenges in using history and philosophy of science in teaching. *Sci. Educ.* **2015**, *24*, 349–385. [CrossRef]

physics MDPI

Article

Phenomena and Principles: Presenting Quantum Physics in a High School Curriculum

Efraim Yehuda Weissman [1,2], Avraham Merzel [1,*], Nadav Katz [3] and Igal Galili [4]

1 The Seymour Fox School of Education, The Hebrew University of Jerusalem, Mount Scopus, Jerusalem 9190501, Israel
2 Department of Mathematics, Jerusalem College of Technology, HaVaad Haleumi 21, Jerusalem 9372115, Israel
3 The Racah Institute of Physics, The Hebrew University of Jerusalem, Edmond J. Safra Campus, Jerusalem 9190401, Israel
4 The Department of Science Education, The Hebrew University of Jerusalem, Edmond J. Safra Campus, Jerusalem 9190401, Israel
* Correspondence: avraham.merzel@mail.huji.ac.il

Abstract: The goal of teaching quantum physics (QP) in high school is a problematic and highly turbulent area of divergent views, curricula studies, and claims. The innovative curricular approach of discipline-culture (DC) suggests a way of overcoming its significant difficulties. It suggests presenting QP as a fundamental theory structured in terms of the *nucleus*, *body*, and *periphery*. Applying this perspective in our study, we interviewed nine experts with respect to their view of how the *nucleus* of QP should be presented to high-school students. With the different viewpoints of the core essentials in hand, we compiled the *nucleus* of the QP. We also examined this subject using nine introductory university textbooks that might suit high school students and considered their coherence and suitability with regard to the specified *nucleus*. We found some confusion regarding the status of theoretical items: some fundamental principles, as perceived in the eyes of the experts, are presented as phenomena. Not only does this mismatch represent a special barrier for both the teachers and students to understand QP, it promotes an inadequate image of QP as well as a distorted view of the nature of science. Finally, we offer a framework for a DC-based QP curriculum free of the noted deficiencies.

Keywords: teaching; quantum physics; discipline-culture; curriculum design

1. Introduction

Quantum physics (QP) exemplifies a central pillar of 20th century physics, and it represents a foundational stone for 21st century technological innovations. However, teaching QP at the high-school level is challenging, and, consequently, it is limited in many countries. The first steps of teaching-learning quantum mechanics are well-represented in a collection of articles [1]. This collection showed different attempts and aspects are still missing one inclusive perspective in the way to represent quantum theory. Recent years have seen attempts to add QP contents; however, the educational literature reveals a highly dissatisfying situation in schools around the world. Teaching QP is usually superficial and oversimplified, often emphasizing phenomena while barely touching on principles and fundamental theory [2,3]. For example, in 2016, the QP curriculum in the Netherlands depicted the phenomena of tunneling, the potential well, and the two-slit experiment as representing QP [4]. An experiment demonstrating the interference of a single photon was specifically developed for this program [5], but a discussion of the wave-particle dualism in this context was missing. The learning assessment tools emphasized the phenomenon, practically without stating the principle of superposition and its impact on measurement [6].

Teaching QP in high schools usually follows a historical approach and because of time constraints, it was abandoned in an early stage of history [3,7]. Although the role of the history of science is widely recognized [8–10], in the case of QP, such an approach can be problematic and sometimes unfeasible [11]. In fact, both a historical and a phenomenological approach fail to adequately represent fundamental theory as a basis of the curriculum. In such approaches, pupils are often presented separate pieces of the theory without a holistic view. Many curricula discuss Plank's explanation of black body radiation, the photoelectric effect, and Bohr's model of hydrogen, which is only semi-classical [2], but they often miss more substantial QP content, including problem solving. Note that the importance of these topics is mainly historical and they have little importance for understanding QP topics [3]. Teaching QP using a historical approach may cause students to believe that QP is nothing more than a certain modification to classical physics (CP), whereas it is actually an essentially different fundamental theory [12]. QP should not be taught by ignoring CP, and the differences between the theories should not be blurred. Teaching must emphasize the departure from classical logic [13,14], avoid oversimplification [15], and be extra careful in using classical analogies [11], which are often inaccurate and misleading. QP should be studied as a theory with a specific set of logic and laws, which require a deep understanding [16].

Another challenge in teaching QP in high school is the level of the mathematic formalism required. It is well above the standard math curriculum, surpassing most pupil's knowledge [2,17]. Visualization is often applied to overcome this deficiency [18,19]. Another approach in this direction involves simulations in order to illustrate the principles of QP, e.g., taking into account the Mach–Zehnder interferometer [16,20]. It should be recognized, however, that the transition to a phenomenological description excludes the possibility of using quantitative questions when teaching physics. This is a significant impediment in presenting physics as essentially a quantitative "exact science". Moreover, teaching QP non-quantitatively is in contrast with the previously taught domains of mechanics and electricity, where a computational approach is the standard method. Instead, students may perceive QP as a collection of stories, leading to an inadequate understanding [21]. In Germany, there was an attempt to introduce mathematical equations (aimed for 13th grade) that provide eigenvalues for simple differential operators [16]. In Italy, teachers tried using Feynman's path integrals, also addressing 13th graders, who are more mature than regular high-school students in other countries that have twelve-year schools [22]. The need to find a mathematical structure suitable for high-school students, which would allow at least some calculations in QP, was evident. Such a mathematical structure should represent quantum theory holistically, beyond a few simplified solutions. With this goal in mind, use of Dirac notations has been suggested as suitable for both computational and conceptual purposes [23].

Curricular designers seek a curriculum that addresses QP misunderstandings, resolving a confrontation with intuition using a more advanced teaching method. Several studies depicted the pertinent misconceptions of university (e.g., [24–26]) and secondary school students [2]. The alternative understandings of the Copenhagen interpretation have been investigated. For example, students may perceive an electron as a cloud in space [26]. However, beyond the required remedy of correcting misconceptions and alternative understandings, a new curriculum should present QP as a fundamental theory, and discuss its principles and applications.

An appropriate approach to building such a curriculum is to consider it as a discipline-culture (DC) [15,27–29]. The DC paradigm suggests presenting QP as a fundamental theory structured in a *nucleus*, *body*, and *periphery*. The *nucleus* includes basic principles and concepts, the *body* includes phenomena and applications, and the *periphery* includes the alternative physical accounts and the limits of validity. Thus, for example, in the classical mechanics picture, the *nucleus* includes Newton's laws, the *body* includes Kepler's laws, the associated subdued elements, solved problems, and applications, whereas the *periphery* includes alternative understandings, principles, including those of Aristotelian physics, relativity, and quantum theories [30]. The *periphery* is important because in order to understand what is correct, it is necessary to contrast it with the corresponding alternative [31]. This picture makes explicit the idea of historical dynamics, the transition from one theory to another, and from one *nucleus* to another [28,32].

The advantage of this approach lies in its holistic presentation. It emphasizes the basic principles of fundamental theory and connects them to their derived phenomena and experiments. We have applied this approach to construct a short course on QP for high school [23,33], which emphasized teaching an existing theory and not how it was historically developed. The principles and their illustrative experiments became more prominent. The *periphery* re-emphasized the principles and positioned the theory (as a whole) in relation to other fundamental theories such as classical mechanics. This approach tried to fit QP to the very limited curricular time-frame allotted by the contemporary school curriculum for modern physics. This constraint dictated a very different curriculum policy as a feasible and still meaningful alternative.

Here, the aim was to refine some aspects of this approach to QP teaching, an approach that elicits and emphasizes principles and relates them to their applications rather than just mentioning them, by explaining such applications and related phenomena, which often occur in the context of teaching. There was an established need for practitioners [34] to be provided with a suitable teaching resource to rely on, learn, provide examples, and acquire competence. Equally important is to identify the principles of QP in parallel with other physics disciplines taught at school. In a way, this attempt is parallel to a similar attempt to teach classical mechanics [30]. There was a curricular implication that changed the regular emphasis with respect to the energy/momentum conservation laws of the *body* of knowledge, as derived from Newton's laws in the *nucleus*. In other words, the division into the *nucleus* and *body* implies certain ideological assumptions to be explicit in teaching: stating the basic principles and what is derived from them, specifically, how can a *nucleus* and a *body* be defined. Consequently, the picture for the students becomes transparent. In accordance, we now specify our research questions.

Here, first, we specify the *nucleus* in the context of QP to be taught in high school, which is not being explored sufficiently in an educational context. Therefore, we explored the views of pertinent experts, physicists directly dealing with QP research. Second, with the *nucleus* initially defined, we turned to the available textbooks for secondary education that can help in this regard.

The obtained answers to these research questions will lead us closer to a better understanding of how we can construct an appropriate DC-structured, QP curriculum for high school, or at least, present an essential step towards an option of this kind. Although there may probably be more than one solution, we will present a detailed proposal for such a curriculum. The effect of actually applying such a curriculum (an intervention study) will be presented elsewhere.

2. Methodology

2.1. Determining the Nucleus via Interviews with Experts

To elicit the QP *nucleus* applicable for high-school teaching, we first interviewed a sample of nine experts (Table 1). Other studies [35] showed that practitioners from different fields in physics have different perspectives on QP; therefore, it was important for us that the chosen experts were physicists engaged directly in the study and teaching of QP. Three of them teach or taught QP courses at the university level where they carry out their research. One was an expert in the philosophy of science with an emphasis on QP. To avoid possible institutional bias, experts from two research universities were selected, including an emeritus scholar and other younger researchers at the beginning of their careers, including a doctoral student with teaching experience. Although they were experts in the subject matter, all but one were very distant from teaching physics in school. Unlike other studies that often considered items of knowledge associated with all parts of DC theory and mostly to the *body* of knowledge [36], we conducted our interviews in order to identify the important concepts of the *nucleus*, revealing the fundamental meaning of QP as a theory. For the purpose of identification, personal interviews could better serve our goals.

Table 1. The interviewees and their domains of expertise.

No.	Position	Seniority (or Years)	Area of Expertise
1	Ph.D. student (physics)	Ph.D. Student	Quantum communication and entanglement
2	Physicist	4	Nuclear and hadronic physics
3	Physicist	40	High energy and strings theory, Lecturer (QP)
4	Physicist	12	Quantum entanglement, Lecturer QP
5	Physicist	36	Condensed matter; Lecturer QP
6	Physicist	14	Quantum coherence
7	Physicist	1	Nuclear astrophysics
8	Physicist	7	Nonlinear quantum optics
9	Philosopher of science	11	Philosophy of physics, classical and quantum statistical mechanics

First five interviews (Table 1) were conducted face-to-face and lasted about an hour each, in the format of semi-structured interviews (e.g., [37]). The rest were conducted at a distance. The experts were presented with the DC approach, and as an opening question, they were asked: "What do you consider as the fundamental QP principles, given that we want to teach them in high school?".

After the experts provided the concepts and principles that they considered as important, they were asked to refine them in order to see the hierarchy of the principles and their meanings. For example, after pointing to "duality" as an important component of the *nucleus*, we asked them to elaborate more about its meaning. Sometimes we asked about the relationship of certain concepts even if they were not mentioned by the interviewee. The duration of the interviews could vary according to the ability and interest shown by the interviewee to the claims made by other colleagues.

The experts defined the items they associate with the *nucleus*, exemplified by *body* items that correspond to the mentioned components of the *nucleus*, and sometimes they also referred to the *periphery*. The interviews were transcribed, and a thematic analysis was conducted: Because DC was presented to them explicitly, it was not a complicated task to analyze the experts' words and to categorize them in triadic terms. After the first author performed a preliminary categorization, the second author joined in its validation through a discussion and full agreement was easily reached. Only some of the data used in this report focused on the *nucleus*. Experts' examples and illustrations (associated with the *body* or *periphery*) are not presented here. Since each concept can have many illustrations for teaching [34], a Delphi study [36] would be a better way to study them.

2.2. *QP Textbooks and the Nucleus*

A sample of nine textbooks on "modern physics" was examined (Table 2). These textbooks, intended for colleges, or as service courses for engineering students, were examined because they should be more suitable for the high-school level, compared with "quantum physics" textbooks aimed at university physics students. Some of the textbooks considered contain much more advanced topics (e.g., nuclear physics, elementary particles), presented mainly phenomenologically with minimal formalism. Such parts, less relevant to high school are important for us in order to define the scope and limitations of our teaching at school, particularly in a DC perspective that connects various phenomena to basic principles. For the purpose of diversity, the sample included both new and old editions of textbooks. The textbooks were examined according to the saturation method for qualitative studies, in which the process ends when adding more items seems not to add new (qualitative) information [38].

Table 2. List of the physics textbooks examined.

(No) Author(s) (Year of Edition)		Title
(1)	Beiser (2003 [39])	*Concepts of Modern Physics*
(2)	Weidner and Sells (1973 [40])	*Elementary Modern Physics*
(3)	Serway, Moses, and Moyer (2005 [41])	*Modern Physics*
(4)	Krane (1983 [42])	*Modern Physics*
(5)	Tipler and Llewellyn (2008 [43])	*Modern Physics*
(6)	Thornton and Rex (2013 [44])	*Modern Physics for Scientists and Engineers*
(7)	Nolan (2014 [45])	*Fundamentals of Modern Physics*
(8)	Noce, Ed. (2020 [46])	*Modern Physics; A Critical Approach*
(9)	Halliday, Resnick, Walker, and Taylor (2012 [47])	*Understanding Modern Physics* (Hebrew translation)

The assumption in looking for fundamental principles is that they are related to numerous items, experiments, and phenomena. The principles had to be referred to as such, and not as phenomena. The analysis was twofold, both quantitative and qualitative.

Within the quantitative analysis, we counted the occurrences of keywords of the *nucleus*, as it was defined from the first part of the study (the interviews). Most of the textbooks were available in digital form. They were searched by digital counting. First textbook (Table 2) was searched manually, using the index. Within the qualitative analysis, we examined how each item of the *nucleus* was referred to. The first and second authors categorized such connections as either a principle or phenomenon. Validation was carried out by comparing categorizations with a short discussion to reach full agreement. Similarly, conceptual definitions, explanations, and examples (including problems) were analyzed, as well as connections between the principles.

Another part of the analysis focused on the book headings (chapters or sub-chapters) and their relationship to the *nucleus*, *body*, or *periphery*.

3. Findings

3.1. Experts on the Nucleus of Quantum Physics

We obtained a significant amount of data from the experts regarding the QP *nucleus*. However, four of the experts (Nos. 1, 3, 6, and 9, see Table 1) hesitated somewhat in defining the *nucleus*. For example, expert No. 6 said: "Quantum physics does not have an agreed upon ontology... As in the parable of the blind examiners and the elephant, each sees quantum physics as something different" (Figure 1).

Figure 1. The parable of the blind and the elephant [48] (p. 89).

Two related difficulties can be seen here. One is the lack of consensus regarding the QP ontology and different controversial views. This revealed the second challenge of a pedagogical nature, namely, the possible approaches to teach QP and its principles, which may pose the question of whether defining the *nucleus* of QP is practical. Expert No. 9 suggested: "Students should be presented with the fact that, on the one hand, quantum physics does not have an agreed-upon worldview, but on the other hand, quantum physics possesses an excellent ability to predict results with an accuracy of 10 digits after the decimal point".

3.1.1. The Items

The wave-particle duality, or the wavity as a property of a particle, is among the central principles that arose in all interviews. For example, expert No.3 said: "Axiomatically, there is a wave function that describes a particle".

Next, we refer to the meaning ascribed to this feature. The experts associated wavity with another principle they considered central: the principle of superposition. For example, expert No.9 stated: "The first principle is that nature includes state superposition. A state can include several probable options".

All experts mentioned the probabilistic aspect of the measurement results, for example, expert No.2 stated: "One of the fundamental principles is the probabilistic explanation of the wave function".

Other principles associated with the nucleus are non-locality and entanglement (experts Nos. 1, 8, and 9). Two of the experts (Nos. 1 and 8) emphasized that only when entanglement is present and realized, can one say that the system is "quantum", i.e., specific of QP. Expert No.1 indicated that quantum nature is essentially different, irreducible to the classical predecessors: " Waves we know, particles we know, but EPR doesn't have a classic analogue".

The EPR (Einstein–Podolsky–Rosen) experiment [49] deals with entangled particles: two particles produced together and comprising a single system. The experiment states the immediate effect of measuring one particle on the state of the other. Therefore, this experiment (realized in the 1980s [50]) demonstrated non-locality in the system described by the wave function unifying the objects. This way, "non-locality" entered into physics as a feature of reality. It implied the relationship between two distant measurements of the quantum objects in a single quantum state, which cannot be explained by any physical interaction between the parts of the system, and the interaction at a finite speed. Indeed, we saw the experts who consider non-locality and entanglement as an essential part of the *nucleus* of the QP as a fundamental theory.

Some experts mentioned Heisenberg's uncertainty principle (experts Nos. 2 and 4) and quantization of energy levels (experts Nos. 2, 3, and 7). For example, expert No. 2 mentioned: "The basic principles are duality, uncertainty, and discrete energy values".

Another principle, mentioned by two experts (Nos. 2 and 7), is the division of elementary particles into bosons and fermions. There are two kinds of indistinguishable particles with respect to their coexistence: the fermions, which obey the Pauli exclusion principle, and bosons, which are not committed to it (expert No. 2): "Fermions and bosons; fermions have the Pauli principle, as opposed to the bosons, which 'want' to be in the same state".

That is, the principles of central importance mentioned by the participants were as follows: duality, superposition, probability in measurement, entanglement, Heisenberg's uncertainty, quantization, as well as the division into bosons and fermions and their properties.

3.1.2. Scrutinizing with Respect to the *Nucleus*

In this Section, we consider what the experts meant when they referred to the *nucleus* concepts. In our opinion, duality was the gateway. When dealing with particles, e.g., an electron, they stated that the novelty lies in also considering the electron as a wave, i.e., while characterizing an electron, they added properties that exhibit wave phenomena. To furnish this idea and to clarify the meaning of "wavity" for matter, we asked the experts directly. In response, all the experts linked wavity to the principle of superposition. They ascribed to it the essence of quantum duality. For example, expert No. 9 said: "Superposition is the most fundamental, in quantum theory in general and in wavity in particular."

Just as a wave is a non-local phenomenon, i.e., it exists in different places and at different amounts, the quantum particle is also in a superposition of locations, as described by a wave function. In the words of expert No. 8: "Wave function [of a particle] consists of a collection of probabilities at each point, a superposition of several likelihoods. For physicists, it is more interesting to talk about discrete states and a superposition of any two states."

This account was generalized to the superposition of the states of other properties such as momentum states, the states of polarization, spin, and others. Expert 7 confirmed this explanation as a "legitimate interpretation". Another expert (No. 6) mentioned that the term "wave" in this context signifies a physical quantity having more than one value: " . . . wave function is superposition; it presents an extension of meaning that occurred in English and German. It is in a sense a borrowed semantic, not physical".

Expert No. 1 clarified the meaning of superposition as follows: "Waves also have superposition, but the novelty here [in QP] is the probability... The probabilistic features have no a classical analogue".

Thus, the important feature is probabilistic superposition, as expressed by the collapse of the wave function in a measurement. This is in contrast to the superposition in a classical case, e.g., in string, where several frequencies and amplitudes are simultaneous. Expert No. 7 stated: "As long as there is no measurement, there is a superposition. We cannot measure superposition. When measured, the state collapses, and if we do it [the measurement] many times, we get statistics [and the idea] of what the initial state was".

That is, according to the Copenhagen interpretation originally proposed by Max Born, quantum superposition means that before a measurement, a state consists of several eigenstates. A measurement produces a certain result, and the wave function collapses. Numerous identical measurements provide results initially described by the wave function.

Regarding measurement and collapse, some experts (Nos. 2, 7, 8, and 9) warned against the understanding of "turning a wave into a particle" (No. 7): "I wouldn't express it as measurement that turns a wave into a particle... It remains a wave function..."

As for Heisenberg's uncertainty and quantization, experts were not unanimous. Expert No. 6 stated: "Quantization, tunneling, and uncertainty are not fundamental principles, but the phenomena arise from fundamental principles, and they can be derived from these principles".

Other experts (Nos. 1 and 5) have also argued that quantization can be a phenomenon in classical waves and that it is not a quantum principle. Another expert (No. 2), to whom these words were presented, addressed the pedagogical aspect of this point: "Even if something derives from the core, it could not always be shown to the pupils immediately... In classical physics, we teach conservation of momentum, but only later, it derives from the Noether theorem".

In other words, if one cannot derive a phenomenon as implied from a principle, perhaps a school teacher can treat it as a principle.

Regarding the meaning of Heisenberg's principle of uncertainty, expert No. 2 commented: "The uncertainty principle is erroneously presented using Heisenberg's microscope... The "uncertainty" is related to a conscious viewer, and this is a problem".

It should be emphasized that the Heisenberg principle refers to the nature of things, rather than depending on the observer.

3.1.3. Interim Summary of the *Nucleus*

There is a consensus among the experts regarding the centrality of the superposition principle in QP. This principle possesses probabilistic meaning, as evident in the collapse of the wave function in measurement. Beyond that, there was some agreement relating wavity to superposition. This approach seems to be suitable for high-school teaching and it should be presented as the greatest conceptual innovation of QP. The wavity of matter constitutes the gateway (historical and conceptual) to QP. Looking at wavity as a superposition of states establishes a unified view essentially different from the classical view. In contrast, the concept of quantization (possessing discrete, non-continuous values) appears to be suitable for the *body*, rather than for the *nucleus*. There are several reasons for that:

A. Methodological: Several experts explicitly argued that quantization represents a phenomenon and not a principle. In addition, the phenomenon of quantization is also observed in CP.
B. Conceptual: Quantization derives from the solution of the wave equation. It can be illustrated by standing waves. Quantization cannot be a fundamental principle in QP, since it characterizes all waves, both classical and quantum.

C. Pedagogical: Pupils encounter quantization while studying about the Bohr model, which is semi classical; it belongs to the *periphery* of QP. Considering it as a fundamental principle is misleading regarding the essential features of QP.

The Heisenberg uncertainty principle stems from a more basic principle (Non-commuting nature of certain observables such as the momentum and coordinate of a particle), which is beyond high-school students' mathematical level. However, the Heisenberg uncertainty principle is related to the state of superposition, determining the kind of states that can be superposed. Therefore, in high school, this state of matter can be presented as claiming the existence of pairs of physical quantities (e.g., position and momentum, spin and polarization in different directions), for which eigenstates cannot coexist in a superposition. In this sense, uncertainty can appear as a principle in teaching and can be referred to with regard to the *nucleus*. Presented as an inequality relating to the accuracies of location and momentum (or life time and energy), it may be used as a phenomenological constraint rather than a basic principle.

We wish to note that the experts did not explicitly mention the complementarity principle. However, they talked about uncertainty and duality in relation to complementarity. Some of them had reservations regarding conceptualizing measurement (and the related collapse) as a transition of the quantum objects from a wave to a particle.

Regarding non-locality and entanglement, it was associated with the *nucleus*. To establish a logical sequence of teaching the *nucleus*, it should be preceded with such concepts as quantum state (eigenstate) and superposition of states. Quantum non-locality could be illustrated in the interference of a single photon with two paths of transition available, two slits experiment or the Mach–Zehnder interferometer (both affiliated to the *body* in the quantum theory). Considering these cases can strengthen students' understanding of superposition and non-locality, also demonstrated by the realization of the EPR scenario as a paradigmatic experiment (In the advanced presentation of QP, the EPR experiment belongs to the *body* knowledge. Yet, in the introductory course, which only touch on non-locality, EPR can be sent to the *periphery*, posing a 'paradox' and promising the students its resolution in their further learning of Quantum theory).

Another item of the nucleus further expands the theory to many particles. After the principle of the indistinguishability of elementary particles in QP, the claim of dividing the micro-world particles into bosons and fermions may follow Pauli's principle for fermions and the possible coexistence of bosons in the same state. The principles of the *nucleus* can be associated with the phenomena of the *body* knowledge—the periodical table of elements for fermions and the laser for bosons.

3.2. *Analysis of Textbooks*

We will first address the textbooks from a more general perspective, followed by how the books refer to the concepts of the QP *nucleus*.

3.2.1. Blurring the Structure of the *Nucleus* and *Periphery*

All the textbooks of the sample begin by presenting special relativity theory. At the introductory level, QP is not relativistic. Thus, the two fundamental theories are essentially different. This implies that their *nuclei* are located in each other's *periphery* [27,28]. In fact, the two theories are combined under the roof of Modern Physics developed at the beginning of the 20th century. Importantly, however, if the curriculum does not deal with the status of a fundamental theory in physics and there is no such theory called Modern Physics, the inclusion of both theories in a text may confuse both the students and teachers regarding the boundaries of the QP *nucleus*.

Modern Physics texts often follow a historical narrative. In it, Planck's account of the black body radiation, Einstein's account of the photoelectric effect, and the Bohr model of an atom follow each other in a line. The Bohr model is semi-classical and belongs to the *periphery* of QP. This historical evolution in no way represents QP as a theory. For instance, in Krane's book [42] Planck's and Einstein's explanations are sub-headed under

"The quantum theory of...". Certainly, this approach represents neither the *nucleus* of the theory nor its *periphery*. The presented history is also inaccurate, since the Bohr model is located after dealing with de Broglie waves. Using de Broglie waves in the Bohr model was heuristic, and it does not explain wavity in the quantum sense as was discussed above.

A review of chapter headings and their subsections reveals a problem when addressing principles (*nucleus*) versus phenomena (*body*), let alone the *periphery*. For example, in Thornton and Rex's text [44], there is a chapter providing a historical introduction, a chapter on Bohr's atom, and a chapter (i.e., with the same status) on the wavity of matter, its meaning, the uncertainty principle, and more. Within the latter chapter, wavity is not associated with a superposition of states. There is no section that defines the basic principles of QP; however, several sections deal with phenomena (scattering, "a particle in a box"). The basic principles do not appear in titles, which would indicate their centrality. Other books in the sample are similar in this respect. Not identifying a *periphery* eliminates any contrast with the *nucleus*, blurring the meaning of different historical discoveries and missing the major principles of QP, which becomes clear by explicitly identifying the *nucleus*, *body*, and *periphery*.

3.2.2. Lack of Stating the Basic Principles

We found that the textbooks examined did not state the basic principles of QP. For example, in one of the books [43], in the chapter *The wavelike properties of particles*, the topic of particle-waves is discussed but no principle was stated, nor were there any fundamental interpretations of wavity in connection to the superposition of states. This makes the claim regarding the wave function totally unclear regarding its connection to a wave (what wave? why a wave?). The statement "$|\Psi|^2$ is proportional to the probability of detecting an electron in a unit volume" [43] (p. 204) appears arbitrary and is not explained.

Another example comes from Nolan [45]. This book presents and postulates (a sort of *nucleus*) both in the context of special relativity and in the context of Bohr's model of the atom. In light of this, the lack of such important definitions, which are central to QP, is very prominent.

3.2.3. Treating Principles as Phenomena

Another observed feature of the presentations in the sampled books is reference to principles as phenomena. Such a reference is made to superposition, the Heisenberg uncertainty principle, and wavity. For instance, wavity is referred to as the ability of an electron to perform diffraction and interference [43] (p. 187), as if in a classical sense. The meaning by addressing the superposition of different states (which does not exist in CP) is totally missed.

1. The same is true in Krane's textbook [42], that is, wavity is treated in the context of de Broglie's hypothesis, the focus is on diffraction and interference (p. 104), and there is no reference to the superposition of states. Later (p. 110), the double slit experiment is described with an examination of the slit in which an electron passes; however, there is no mention of wave function collapse during a measurement. The description mentions that it is related to particle-wise behavior, and if it is a particle, it loses its wavity (the complementarity principle). In this case, the phenomenon is indeed explained by a principle (the *body* is explained by the *nucleus*); however, a few deficiencies can be mentioned: The measurement not only revealed the particle property. It caused the collapse of the wave function from a two-state superposition to a single state (particle-wise). This is the same for a measurement made in the slits and a measurement made on the screen. This perspective is absent in the textbook.
2. The principle of complementarity is not explained here. Here, the principle is not distinguished from phenomena. The title *Through Which Slit Does the Particle Pass?* focuses on the phenomenon. The principle is not mentioned.
3. The principle of complementarity rarely appears in the book. It was referred to in the claim of duality; the wave and particle properties complement each other. The only

other case is where it is mentioned in relation to Bohr's contribution to QP, too little for the central principle of this theory.

Recall that the interviewed experts did not mention the complementarity principle as among the fundamental principles of QP. Complementarity has several faces, among which is the uncertainty principle. In the textbooks, complementarity is presented in the aspect of wave-particle duality, and the transition from wave to particle at the moment of measurement. This is unlike the experts who referred to the wave function collapse in the measurement but did not relate that to complementarity. This point, the impact of measurement on a quantum state, seems to us as important and was frequently missing.

Thornton and Rex's book [44] has more emphasis on principles. For example, there are two pages (pp. 191–192) under the title *Probability, Wave Functions, and the Copenhagen Interpretation*, but there is still no comparison between the Copenhagen interpretation and any other. The discussion mainly addresses the historical evolution of the QP theory and less on its holistic image.

Below, in Sections 3.2.4–3.2.7, we present how the books deal with some major components of the QP *nucleus*.

3.2.4. Superposition

In seven of the books chosen, the term "superposition" appears exclusively in addressing the interference between wave functions causing an interference pattern (in one of them, the word does not appear even once in the index). The definition of superposition as a fundamental principle that allows a particle to be simultaneously in several different states in superposition—the core feature of QP—is lacking.

In textbook [42], the term "superposition" appears only four times. The Bohr model appears after particle waves. The Bohr model is semi-classical and deterministic. It determines well-defined orbits and the energies of the electron in the hydrogen atom. It ignores superposition and is certainly inconsistent with the probabilistic aspect of quantum superposition. Its introduction after "matter waves" positions it as an advanced quantum model. This is not the case, mainly because it lacks quantum wavity and superposition as its essence.

In another textbook [43], the term superposition addresses classical waves (5 times), but its use in a quantum context (3 times) lacks a substantive explanation of superposition as a unique principle that allows several states to exist simultaneously. The phenomenon of the wave function collapse in measurement is absent (the term "collapse" does not appear). The term superposition appears in the context of interference and diffraction but never regarding the ability to coexist in multiple states. Nonetheless, superposition is mentioned in the more advanced topics, such as the Bose–Einstein condensate and elementary particles, but without an explanation presenting it as a principle.

Another example comes from Beiser's textbook [39], where superposition appears in two locations (four times). It addresses classical waves (p. 55) as a principle of combined waves. The other case is when considering the Schrödinger equation (pp. 169–170). It is used to present the mathematical aspects of the linearity of the equation, allowing its solutions to be combined. Superposition does not appear as a principle of the possibility of co-existing quantum states (e.g., in considering non-locality).

Only in book [46] does the superposition appear as a physical principle that addresses the combining of two states. In other textbooks in the sample, superposition is mentioned only as a mathematical feature, and not as a central and unique physical principle with probabilistic meaning.

3.2.5. Measurement and the Collapse of the Wave Function

The probabilistic aspect of QP is an established item of the *nucleus* (according to experts' interviews). It is multiply referred to in all the textbooks. To detect problems, we will address two topics linked to this concept: measurement and collapse of the wave function.

In eight of the textbooks, the term "collapse" does not appear at all. The books ignore it, and also lacked any other interpretation of measurement, similar to the "many worlds" interpretation. The issue of the essence of QP is lacking. In addressing measurement, the textbooks did not dedicate chapters or sections to define measurement as projecting to a new state. Seemingly, in the context of measurement, probabilistic results appear as a phenomenon: using the wave function, one calculates the probability of certain results in measurement. The books do not state the fundamental difference between the quantum and classical probabilities. The former presents the intrinsic feature of the QP, its *nucleus*, and differs from classical probability in statistical mechanics, technically imposed by the lack of information regarding the states of all particles. For example, Serway, Moses, and Moyer [41] (p. 192) refer to the wave function as a probability function describing the Born Interpretation of the phenomenon. The matter is not related to a state's superposition, a subject of intrusion in the course of measurement, and a probabilistic transfer to a single state. In other words, the *nucleus* is not contrasted with the *periphery* (quantum vs. classical physics). This terminology could be useful for pointing to the principles on which other ideas are based.

Similarly, Krane's book [42] (p. 110), when describing the measurement on the screen in the double slit experiment, does not mention the collapse of the electron's wave function and its probabilistic nature being valid for the flux of many particles. A missed opportunity was to state that the same collapse occurs in the detector instead of on the screen when the former was added next to the slits.

The book [46] differs from the rest. It relates to collapse (26 times), the interpretation of many worlds (5 times), and emphasizes the uniqueness of quantum measurement [46] (p. 8–39): "The postulate that translates operationally this view [the Copenhagen interpretation] is the one that takes into account the so-called collapse of the wave function. According to this hypothesis, when a quantum system is measured it interacts with a classical instrument, which forces the measured quantity of the quantum system to assume only one of the possible eigenvalues for that observable, so that the wave function instantly changes, i.e., collapses, in the eigenfunction associated with the selected eigenvalue."

All together, most of the textbooks lack a discussion on the meaning of measurement in QP as a unique feature through which we obtain knowledge of the micro-world.

3.2.6. Heisenberg's Uncertainty Principle

All the books refer to Heisenberg's uncertainty principle, and it does not suffer from a lack of emphasis (more than 100 references). At the same time, there are a number of problematic aspects in this regard. It has been often treated as a phenomenon and not as a principle. For example, [39] proclaims: "This principle... is one of the most significant physical laws"; yet it states [39] (p. 108): "It is impossible to know both the exact position and exact momentum of an object at the same time".

It appears in a separate line, highlighted as a definition of a principle. In a way, avoiding an explicit reference to uncertainty as an intrinsic feature in nature may be understood by learners as a difficulty, insurmountable, but it does not exclude the existence of the exact values. Next, one reads (p. 112): "A measurement establishes the position of a proton with an accuracy of $\pm 10^{-11}$ m. Find uncertainty in the proton's position 1s later. Assume $v << c$".

Presented as a phenomenon, the difficulty of *our ability* to know is emphasized. A similar motif appears in the explanation of "Heisenberg's microscope" (p. 113). The uncertainty in momentum is caused by the photon that measured the position. This explanation is deficient. As was mentioned in the experts' views, it refers to the uncertainty in our knowledge (the observers who need a photon) rather than a fundamental limitation. Next, on pp. 114–116, in addressing the minimal kinetic energy, the uncertainty appears independent of the observer. Thus, the two references to Heisenberg's principle are sometimes mixed: as the limitation introduced by observation and as intrinsic feature of reality. This blurs the difference between principle and phenomenon being potentially confusing at any level of teaching, especially in high school.

The uncertainty is often stated in the unclear status as our inability to be precise [41] (p. 174): "If a measurement of position is made with precision Δx, and a simultaneous measurement of momentum in the x direction is made with precision Δp_x, then the product of the two uncertainties can never be smaller than $\frac{\hbar}{2}$. That is, $\Delta p_x \Delta x \geq \frac{\hbar}{2}$." Here, $\hbar$ is the reduced Planck's constant.

The definition is presented as a highlighted paragraph. The origin of uncertainty is ignored. This fuzziness of its origin is common: our limitation, a phenomenon, is not a principle. It is never mentioned in the context of the superposition of states. In the whole sample, uncertainty is addressed only regarding the momentum-position or time-energy.

Another example we encounter in the definition appears in [42], which starts with (p. 116): *It is not possible to make a simultaneous determination of the position and the momentum of a particle with unlimited precision.*

A formula that links Δx to Δp follows. Yet the author proceeds with emphasis on our ability to be accurate. He states that inaccuracy of Δx implies inaccuracy of momentum and clarifies the essence [42] (p. 116):" The Heisenberg relationships are sometimes called "indeterminacy" rather than "uncertainty" principles, because the idea of uncertainty may suggest an experimental limit that can be reduced by using better equipment or technique".

The emphasis is explicit; it is a matter of principle. The author warns of confusion. Yet, the principle is never stated as implying the absence of being a particle in eigenstates of position and momentum simultaneously.

Thus, the uncertainty principle receives the status of a physical property and not the human inability to be precise, but even in these cases, it does not go beyond addressing the momentum-position and time-energy. In addition, it is often presented as a phenomenon, described by a formula, and in most textbooks, it ignores the eigenstates and the superposition states. In other words, it does not appear as a principle, it never mentions superposition, but rather, it refers to it as a curious phenomenon related to position and momentum, which have a certain relationship.

3.2.7. Entanglement and Non-Locality

The authors perceive entanglement and non-locality as an advanced topic. In seven books, these concepts were not mentioned. Non-locality appeared among the advanced topics in the book [46], in the context of the EPR paradox (p. 8–43). Nolan's text [45] did not address non-locality as a physics feature, but employed it in describing quantum teleportation (six times, pp. 354–355).

In conclusion, our findings indicate that the textbooks suitable for teaching QP in high school do not emphasize enough the principles, often present them as phenomena, and do not provide a set of principles on which the considered theory is built.

4. Discussion and Conclusions

Designing a curriculum for teaching QP at the high-school level is challenging in terms of the conceptual difficulty, the peculiarity of the theory, and the required mathematics (e.g., [35,51]). A special barrier is a lack of a pedagogical tradition to present a uniform theory instead of detached topics. Quantum theory essentially differs from the classical one and is not a refinement of it. An additional difficulty stems from the very limited time reserved for it in the curriculum. We claim that the DC paradigm of physical theory serves as an adequate tool in facing these difficulties. Its advantage is a well-defined hierarchical structure that carefully defines the components of the theory starting from its *nucleus*. This new requirement is demanding because such a tradition is lacking in the context of high-school teaching. We have arrived at a set of conceptual fundamental principles through interviews with experts in QP. We then examined the way in which these principles are expressed in textbooks at the relevant level and extracted additional insights.

In our defining the basic principles feasible for teaching QP in high school, we consider as central the wave-particle duality, with particle wavity defined as the ability of being in a superposition of states. This is consistent with the mathematical claim of linearity from which the superposition principle derives [12,51,52]. The quantum superposition is probabilistic, which makes the particle wavity specific in QP. This nature is manifested in measurements in which the collapse of the wave function reduces a quantum object into a particle-like state. Despite its centrality, we found this discussion extremely rare in the available teaching materials.

Heisenberg's uncertainty principle is commonly known as the limit of the accuracies of two quantities that cannot coexist as exact values. In high school physics, this relationship cannot be derived from the more general claim. We argue for teaching, in addition to the common form of the uncertainty principle, the conceptually equivalent claim with regard to the same pair of quantities prohibiting their simultaneous being in eigenstates. In other words, given that one quantity is in an eigenstate implies the other to be in a superposition of its eigenstates. Such an implication of the uncertainty principle strengthens our qualitative understanding of QP.

We observed that the experts did not mention the complementarity principle. Similarly, it appears to be not popular in the textbooks. (There is a different approach to preserving the centrality of the principle of complementarity [12]). This is an educational phenomenon, given that complementarity is a central philosophical principle of quantum theory. In fact, the complementarity principle (historically related to Niels Bohr) is closely related to Heisenberg's uncertainty principle, and the latter can be seen as a quantitative refinement of the former. Both principles involve measurement in the quantum world. Bohr's complementarity is not quantitative. It might explain its strange absence in introductory textbooks and teaching discourses. Yet, the conceptual relationship between the two principles would be definitely conceptually rewarding and it represents an area that requires an extra effort from physics educators.

4.1. Principles and Phenomena: Teaching the Body as the Nucleus

The division between *nucleus* and *body* is not always unequivocal. In the context of high school, which is given significant limitations of different kinds, some phenomena that actually arise from a fundamental but inaccessible principle can be referred to as a *nucleus*. A prominent example of this is Heisenberg's uncertainty principle, which is not fundamental in the advanced course, but it can be so in high school. Thus, some canonical phenomena, such as quantization, tunneling [53,54], and interference [55] might sometimes be treated as a "concept" or "principle".

In some previous studies, there was a search for key-topics in QP for high school, through various research methods (e.g., [3,35,54]). They categorized their findings to "fundamental principles" (or "concepts"), "examples" (sometimes include "experiments" as a different category), and "applications". Other categories were "philosophical aspects", "atomic theory" (which is sometimes included under "examples") and "mathematical representations". Certainly, what we found and define as the *nucleus* is congruent with their first category, but a refinement according to the triadic framework of the topics they consider as fundamental would show that it includes concepts that we define as the *body* (e.g., quantization, tunneling); and they are silent about the *periphery*. This may be because of the lack of distinction between "what is important to teach in high school QP" and "what is the nucleus of QP". The first, should definitely include aspects of the *body* (examples, applications), therefore canonical phenomena might get a principal status. However, although "duality" is unanimously a fundamental concept, in many cases it is described according to the behavior as a wave or as a particle in different experiments with no connection to superposition, as we mentioned above. That is, the *body* is presented as the *nucleus*. In addition, whereas some experts thought that teaching superposition is too complex (e.g., [54], p. 316), we consider it as the gateway for understanding QP and immanent to high-school QP curriculum. Yet, we do not exclude that there could be some circumstances to make a different choice of *nucleus*.

What is pedagogically important is that the DC approach requires a clear distinction between the *nucleus* and the *body* and *periphery*. In a way, one may put it as distinguishing between principles and the corresponding illustrative phenomena. Anything that derives from basic principles and can demonstrate or explain them can be categorized as the "*body*" knowledge. That includes interference, tunneling, quantized spectra, and more. Such an organization of the curriculum by DC can effectively overcome the challenges of teaching described above, assist pupils [30,33] and teachers [56] in understanding the subject matter and can significantly improve students' perceptions of the nature of science and scientific knowledge [30,57,58].

4.2. *Treating Principles as Phenomena: Teaching the Nucleus as the Body*

Unawareness of the DC structure may blur the boundary between the *nucleus* and the *body* by referring to principles as phenomena. Introductory QP textbooks usually introduce matter waves mainly through the phenomena of interference and diffraction of electrons. However, there is another way by addressing waves through the fundamental definition of the superposition of states [33]. A clear example of the former could be by presenting single photons in an experiment [4]. The alternative instruction, although describing interference, refers to the principle of superposition, and the collapse of the wave function in measurement. Similarly, presenting the uncertainty principle as our "inability" to know focuses on the phenomenon instead of the principle. In the DC approach, the phenomena demonstrate the principles and do not replace them; interference is not wavity. Instead of equating wavity with interference, the teacher states that wavity leads to interference, even if historically the order of the principle-phenomenon could be reversed.

Moreover, stating that the double-slit experiment is a phenomenon not only demonstrates wavity by interference, it also demonstrates a whole set of principles, a superposition of states, the non-locality of an electron that interferes with itself, the role of measurement (the screen as a detector), and the collapse of the wave function. Taken as a phenomenon, this experiment represents several basic principles. This is not obvious for different stakeholders of QP [34], but to appreciate this, one needs to distinguish between principles and their application.

In the DC approach, we try not to present a phenomenon and explain it through a principle, but rather, we claim the main thing, the *nucleus*–the principle, through the *body*–the phenomenon. In that way, we avoid confirming that wavity *is* interference or vice versa, stating that nature behaves according to the *principle* of wavity; hence, we observe a *phenomenon* of interference. We found this teaching suitable for the kind of challenges and severe limitations that we face in teaching QP at school [23,33,58].

We observed that the textbooks do not identify certain knowledge as the *periphery* of QP. Historical introductions often mix the items we identify as the *nucleus* in contrast to others identified as the *periphery*. The image of quantum theory loses clarity, which appears in contrast with the *periphery* and supports the understanding of principles—the *nucleus*. This effect is known, and an understanding of the wrong creates an understanding of the right. If one skips a comparison with the previous knowledge, we reduce the chances that the cognitive processes regarding conceptual change, consequently constructing new knowledge [59,60]. The absence of a comparison deprives the *nucleus* of its status, principles, and consequently, phenomena become indistinguishable.

4.3. *Reference to the Principles at a Lower Level*

Sometimes a principle is treated at a lower level than it actually deserves. Thus, teaching superposition solely as a mathematical feature while discussing interference, without emphasizing that it allows the superposition of states, greatly reduces the centrality of this principle. Likewise, Heisenberg's uncertainty is often referred to as a limitation in the relationship between position and momentum, without regard to the fundamental split among observables with direct implications to the kind of observable superposition states, which devaluates the importance of these fundamental principles.

Construction of a curriculum requires determining the content considered as central, regardless of the accuracy of its form. This should be backed up by theoretical and empirical considerations appropriate for the considered population. It is rather common to believe that QP cannot be taught in high school, and that it is a subject only taught at the university level. In effect, this challenge implies the need for a specific approach, a special perspective, and adapted tools, which are technically and philosophically feasible for this goal [23,36]. We believe that the findings of this study help in that endeavor. Specifically, we seek to bridge the gap between the teaching of QP at high school and higher levels. There is much in common in the *nucleus* (principles), *periphery*, and *body*, even if the way to teach and to construct the content must be different.

4.4. Making a DC Curriculum for High School

All of the above supports the need to make a curriculum based on a *nucleus*, *body*, and *periphery*. One can see that that this program will differ from other programs regarding its structure, content, and in the emphasis provided to various details.

First, because of the short time allocated for teaching, the curriculum should focus on QP and less on its historical introductions such as black body radiation and Bohr's model of the atom [3]. Second, reviewing worldwide curricula reveals that topics such as "Discrete energy levels", "Wave-particle duality", "Interaction between light and matter", "Technical applications", and "Matter waves, quantitative (calculations with the De Broglie wavelength)" are the ones that are taught the most [3]. In these topics, the emphasis is on phenomena much more than the principles of QP. (Even though duality is considered in Stadermann et al. [3] as a fundamental principle, it is often regarded as phenomena like the double slit experiment or the Mach-Zehnder interferometer.) Therefore, we do not claim that teaching the phenomena is not recommended; after all, for the purpose of learning, the abstract should be anchored to the concrete. Nevertheless, the focus and organization of QP curricula should differ.

We begin with a single particle and its state. It can be an eigenstate of a physical observable or a superposition of states. The *nucleus* defines wavity as a superposition whose coefficients possess probabilistic meaning, as realized in the collapse of the wave function in measurement. Heisenberg's uncertainty claims the existence of pairs of physical quantities, such as the position-momentum pair, in which being in an eigenstate implies that the mate is in a state of superposition. This is the *nucleus*.

The *body* includes the double slit experiment for electrons. It clarifies the principles of the *nucleus*. Its account includes interference of the wave functions of a single electron. The collapse of the wave function takes place on a screen, producing a specific position with the probability determined by the wave function. The interference pattern emerges after the screen is hit numerous times. Another case considers the output with a detector placed in one of the slits. The new pattern on the screen is compared with the one without the detector, illustrating the role of measurement, in the slits or on the screen.

The following example considers spin and polarization. The case of three measurements in succession is considered. It appears that each measurement deletes the "memory" of the electron or photon regarding its previous state. The Mach–Zehnder interferometer can follow further by illustrating the principles of superposition and the wave function collapse. As a special example, we used the BB84 quantum encryption protocol.

The items of the *periphery* emphasize the *nucleus*. For instance, we refer to the historical idea that the electron in an atom presents a cloud spread in space (an element of the periphery). Another example is the nature of quantum uncertainty essentially different from the uncertainty in classical statistical physics (an element of the periphery). Another misconception, a particle moving in a wave-like path can be also affiliated to the periphery of QP.

Table 3 illustrates the idea of inserting curricular content along with its tripartite organization into a DC structure. The content expands in the cases of single, double, and multiple particles (bosons and fermions). For each item of the *nucleus*, there are examples from the *body* and alternative conceptions of the *periphery*. We argue that such organization of the curriculum is not only representative but also seems beneficial [33] in the reality of a highly limited number of teaching hours. The corresponding teaching should include active learning (e.g., using worksheets and simulations). Being innovative in creating the content of the new curriculum requires organized and guided training for teachers [56].

Table 3. An example of DC-type of curricular structure of QP in high school. See text for details.

Nucleus	Body	Periphery
State–eigenstate and the principle of the superposition of states; the wave function The wavity of matter and superposition. Probabilistic interpretation and measurement Heisenberg's uncertainty and complementarity principle	Dirac notations The double-slit experiment with electrons Spin and polarization The Stern–Gerlach experiment Mach–Zehnder interferometer The BB84 protocol	Classical state and probability in mechanics and thermodynamics Classical measurement without disturbance Electron as a cloud Classic uncertainty—the lack of knowledge
Entanglement	An experiment to examine Bell's inequality	Hidden variables
Quantum indistinguishability Bosons and fermions	Laser, Pauli's exclusion principle, The Mendeleev periodic table	Particles distinguishable in classical statistics. Unification of matter.

4.5. Dirac Notation

An important part of designing the curriculum is adapting proper mathematical tools of presentation. There is certainly room to combine the teaching of the Schrödinger equation and its solutions, and the teaching of simple differential operators. However, subject to a limited timetable, it is necessary to consider the feasibility of such methods, which may be very mathematical and may disregard important conceptualizations. It appears possible, however, to use a special symbolic language, namely, the formalism of Dirac notations [23], to circumvent the challenge of mathematical representation and efficiently deliver the conceptualizations of QP. Such formalism allows representing much of the contents of Table 3, such as the principles of superposition, measurement, and collapse of the wave function of a few state systems (spin half and one).

An expression such as $|\psi\rangle = \frac{1}{2}|\uparrow\rangle - \frac{\sqrt{3}}{2}|\downarrow\rangle$ denotes the superposition of spin states; it represents a superposition state of several eigenstates and the simplest wave function of two states. It indicates the amplitudes allowing the calculation of probabilities to emerge in a measurement as well as the relative minus sign, which distinguishes this superposition state from another state with identical probabilities. That is, writing the state in such a way allows one to visualize the content often hidden in a complex mathematical form that normally prevents any computational context in high school. In our class, it enabled students to consider numerical problems requiring a simple calculation of probabilities and to deal with important cases involving spin or polarization.

5. Coda and Future Research

In this study, we discussed teaching QP in high school within the new educational paradigm of discipline-culture. In particular, we elaborated on the content of the *nucleus*—the core essential principles and concepts of quantum theory as a fundamental physics theory. We made all of this possible in a form suitable for high-school teaching. Within this perspective, we interviewed a group of experts, who were university researchers in QP, and examined how their ideas regarding the *nucleus* are (and are not) reflected in textbooks relevant for high-school teaching. We presented the idea of a curriculum structure based on the three components of the DC structure. Further research should examine the results of utilizing the DC approach for teaching and its impact on both the students' comprehension of QP and more generally, on their perceptions of the nature of science. Specifically, we are curious regarding the impact of the quantitative aspect of teaching with the use of Dirac notation, which seems very promising. Further research should accompany teachers' training when teaching QP and should elucidate students' difficulties, as well as determine the possibility of teaching additional content such as the Schrödinger equation, available in the same formalism, which was not mentioned as one of the components of the *nucleus*, but whose importance (in the *body*) is understood.

Author Contributions: Conceptualization, E.Y.W., A.M., N.K. and I.G.; Formal analysis, E.Y.W.; Investigation, E.Y.W. and A.M.; Project administration, A.M.; Supervision, N.K., I.G. and A.M.; Writing—original draft, E.Y.W. and A.M.; Writing—review & editing, A.M., N.K. and I.G. All authors have read and agreed to the published version of the manuscript.

Funding: This research received no external funding.

Institutional Review Board Statement: This study has the IRB approval: 2021Y0606.

Informed Consent Statement: All participants participated under informed consent.

Data Availability Statement: The data are available upon request.

Acknowledgments: We wish to thank the experts interviewed for their time and insights.

Conflicts of Interest: The authors declare no conflict of interest.

References

1. Zollman, D. (Ed.) Research on Teaching and Learning Quantum Mechanics. Papers presented at the Annual National Association for Research in Science Teaching (NARST) Meeting, Boston, MA, USA, March 1999. Available online: https://web.phys.ksu.edu/papers/narst/QM_papers.pdf (accessed on 20 September 2022).
2. Krijtenburg-Lewerissa, K.; Pol, H.J.; Brinkman, A.; van Joolingen, W.R. Insights into teaching quantum mechanics in secondary and lower undergraduate education. *Phys. Rev. Phys. Educ. Res.* **2017**, *13*, 010109. [CrossRef]
3. Stadermann, H.K.E.; van den Berg, E.; Goedhart, M.J. Analysis of secondary school quantum physics curricula of 15 different countries: Different perspectives on a challenging topic. *Phys. Rev. Phys. Educ. Res.* **2019**, *15*, 010130. [CrossRef]
4. Van den Berg, E.; Brandt, H.; van Rossum, A.; van der Veen, J. Retention of a double slit single photon interference demonstration of particle-wave duality. *J. Phys. Conf. Ser.* **2021**, *1929*, 012049. [CrossRef]
5. Van den Berg, E.; van Rossum, A.; Grijsen, J.; Pol, H.; van der Veen, J. Teaching particle-wave duality with double-slit single-photon interference in Dutch secondary schools. In *Research and Innovation in Physics Education: Two Sides of the Same Coin*; Guisasola, J., Zuza, K., Eds.; Springer: Cham, Switzerland; pp. 135–143. [CrossRef]
6. Stadermann, H.K.E.; Goedhart, M.J. Secondary school students' views of nature of science in quantum physics. *Int. J. Sci. Educ.* **2020**, *42*, 997–1016. [CrossRef]
7. Bitzenbauer, P. Effect of an introductory quantum physics course using experiments with heralded photons on preuniversity students' conceptions about quantum physics. *Phys. Rev. Phys. Educ. Res.* **2021**, *17*, 020103. [CrossRef]
8. Abd-El-Khalick, F.; Lederman, N.G. The influence of history of science courses on students' views of nature of science. *J. Res. Sci. Teach.* **2000**, *37*, 1057–1095. [CrossRef]
9. Matthews, M.R. *Science Teaching: The Contribution of History and Philosophy of Science*; Routledge/Taylor & Francis Group: New York, NY, USA, 2014. [CrossRef]
10. Solomon, J.; Duveen, J.; Scot, L.; McCarthy, S. Teaching about the nature of science through history: Action research in the classroom. *J. Res. Sci. Teach.* **1992**, *29*, 409–421. [CrossRef]
11. Greca, I.M.; Freire, O., Jr. Does an emphasis on the concept of quantum states enhance students' understanding of quantum mechanics? *Sci. Educ.* **2003**, *12*, 541–557. [CrossRef]
12. Pospiech, G. Uncertainty and complementarity: The heart of quantum physics. *Phys. Educ.* **2000**, *35*, 393–399. [CrossRef]
13. Ireson, G. A multivariate analysis of undergraduate physics students' conceptions of quantum phenomena. *Eur. J. Phys.* **1999**, *20*, 193–199. [CrossRef]
14. Ireson, G. The quantum understanding of pre-university physics students. *Phys. Educ.* **2000**, *35*, 15–21. [CrossRef]
15. Levrini, O.; Bertozzi, E.; Gagliardi, M.; Tomasini, N.G.; Pecori, B.; Tasquier, G.; Galili, I. Meeting the discipline-culture framework of physics knowledge: A teaching experience in Italian secondary school. *Sci. Educ.* **2014**, *23*, 1701–1731. [CrossRef]
16. Müller, R.; Wiesner, H. Teaching quantum mechanics on an introductory level. *Am. J. Phys.* **2002**, *70*, 200–209. [CrossRef]
17. Pospiech, G. Philosophy and quantum mechanics in science teaching. *Sci. Educ.* **2003**, *12*, 559–571. [CrossRef]
18. Escalada, L.T.; Rebello, N.S.; Zollman, D.A. Student explorations of quantum effects in LEDs and luminescent devices. *Phys. Teach.* **2004**, *42*, 173–179. [CrossRef]

19. Zollman, D.A.; Rebello, N.S.; Hogg, K. Quantum mechanics for everyone: Hands-on activities integrated with technology. *Am. J. Phys.* **2002**, *70*, 252–259. [CrossRef]
20. Pereira, A.; Ostermann, F.; Cavalcanti, C. On the use of a virtual Mach–Zehnder interferometer in the teaching of quantum mechanics. *Phys. Educ.* **2009**, *44*, 281–291. [CrossRef]
21. Bøe, M.V.; Henriksen, E.K.; Angell, C. Actual versus implied physics students: How students from traditional physics classrooms related to an innovative approach to quantum physics. *Sci. Educ.* **2018**, *102*, 649–667. [CrossRef]
22. Malgieri, M.; Onorato, P.; De Ambrosis, A. Teaching quantum physics by the sum over paths approach and GeoGebra simulations. *Eur. J. Phys.* **2014**, *35*, 055024. [CrossRef]
23. Pospiech, G.; Merzel, A.; Weissman, E.; Katz, N.; Galili, I.; Zuccarini, G.; Santi, L.; Michelini, M. The role of mathematics in teaching quantum physics at high school. In *Teaching-Learning Contemporary Physics: From Research to Practice*; Jarosievitz, B., Sükösd, C., Eds.; Springer Nature Switzerland: Cham, Switzerland, 2021; pp. 47–70. [CrossRef]
24. Bungum, B.; Angell, C.; Tellefsen, C.W.; Henriksen, E.K. Physics students' understanding of fundamental principles in quantum physics. In *Proceedings of the ESERA 2015 Conference: Science Education Research: Engaging Learners for a Sustainable Future, Helsinki, Finland, 31 August–4 September 2015*; Part 1: Learning Science: Conceptual Understanding; Finlayson, O., Pinto, R., Eds.; University of Helsinki: Helsinki, Finland, 2016; pp. 11–19. Available online: https://www.esera.org/publications/esera-conference-proceedings/esera-2015 (accessed on 20 September 2022).
25. Mashhadi, A.; Woolnough, B. Cognitive mapping of advanced level physics students' conceptions of quantum physics. In *Proceedings of the Annual Meeting of the Singapore Educational Research Association*; Singapore, 22–24 November 1995. Available online: https://files.eric.ed.gov/fulltext/ED414195.pdf (accessed on 20 September 2022).
26. Niedderer, H.; Bethge, T.; Cassens, H. A simplified quantum model: A teaching approach and evaluation of understanding. In *Relating Macroscopic Phenomena to Microscopic Particles. A Central Problem in Secondary Science Education*; Lijnse, P.L., Licht, P., de Vos, W., Waarlo, A.J., Eds.; CD-ß Press: Utrecht, The Netherlands, 1990; pp. 67–80. Available online: https://www.fisme.science.uu.nl/toepassingen/20006/ (accessed on 25 September 2022).
27. Galili, I. Promotion of cultural content knowledge through the use of the history and philosophy of science. *Sci. Educ.* **2012**, *21*, 1283–1316. [CrossRef]
28. Galili, I. *Scientific Knowledge as a Culture: The Pleasure of Understanding*; Springer Nature Switzerland: Cham, Switzerland, 2022. [CrossRef]
29. Tseitlin, M.; Galili, I. Physics teaching in the search for its self: From physics as a discipline to physics as a discipline-culture. *Sci. Educ.* **2005**, *14*, 235–261. [CrossRef]
30. Galili, I.; Goren, E. Summary lecture as a delay organizer of cultural content knowledge: The case of classical mechanics. *Sci. Educ.* **2022**. [CrossRef]
31. Marton, F.; Tsui, A.B.M.; Chik, P.P.M.; Ko, P.Y.; Lo, M.L. *Classroom Discourse and the Space of Learning*; Routledge: New York, NY, USA, 2004. [CrossRef]
32. Kuhn, T.S. *The Structure of Scientific Revolutions*; University of Chicago Press: Chicago, IL, USA, 1970; Available online: https://www.lri.fr/~{}mbl/Stanford/CS477/papers/Kuhn-SSR-2ndEd.pdf (accessed on 20 September 2022).
33. Weissman, E.Y.; Merzel, A.; Katz, N.; Galili, I. Teaching quantum mechanics in high-school—Discipline-Culture approach. *J. Phys. Conf. Ser.* **2019**, *1287*, 012003. [CrossRef]
34. Merzel, A.; Andreotti, E.; Antilla, D.; Bondany, M.; Chiocalo, M.; de Cock, M.; Faletic, S.; Foti, C.; Frans, R.; Gabris, A.; et al. Similarities and Differences between Professions and Countries of Key-Concepts in Teaching Quantum Physics and Their Illustrations. In Proceedings of the GIREP 2022 Conference: Effective Learning in Physics from Contemporary Physics to Remote Settings, Ljubljana, Slovenia, 4–8 July 2022; Poster MDR_68_Merzel. Available online: https://sites.google.com/view/girep2022posters/mdr (accessed on 20 September 2022).
35. Winkler, B.; Bitzenbauer, P.; Meyn, J.-P. Quantum physics $\neq$ quantum physics. A survey of researchers' associations. *Phys. Educ.* **2021**, *56*, 065031. [CrossRef]
36. Krijtenburg-Lewerisa, K.; Andreotti, E.; Antilla, D.; Bondany, M.; Chiocalo, M.; de Cock, M.; Faletic, S.; Foti, C.; Frans, R.; Gabris, A.; et al. Quantum Technology Education. Community-Based Development of the Quantum Concept Inventory. In Proceedings of the GIREP 2022 Conference: Effective Learning in Physics from Contemporary Physics to Remote Settings, Ljubljana, Slovenia, 4–8 July 2022.
37. Savin-Baden, M.; Howell Major, C. *Qualitative Research: The Essential Guide to Theory and Practice*; Routledge: Abingdon, UK, 2013.
38. Saunders, B.; Sim, J.; Kingstone, T.; Baker, S.; Waterfield, J.; Bartlam, B.; Burroughs, H.; Jinks, C. Saturation in qualitative research: Exploring its conceptualization and operationalization. *Qual. Quant.* **2018**, *52*, 1893–1907. [CrossRef]
39. Beiser, A. *Concepts of Modern Physics*; McGraw-Hill Companies, Inc.: New York, NY, USA, 2003. Available online: https://safehandsakola.org/downloads/Physics/Concepts%20of%20Modern%20Physics%20-Arthur%20Beiser.pdf (accessed on 20 September 2022).
40. Weidner, R.T.; Sells, R.L. *Elementary Modern Physics*; Allyn and Bacon, Inc.: Boston, MA, USA, 1973. Available online: https://abru.ac.ir/files/teachers/doc-1632918186.pdf (accessed on 20 September 2022).
41. Serway, R.A.; Moses, C.J.; Moyer, C.A. *Modern Physics*; Brooks/Cole, Thomson Learning: Belmont, CA, USA, 2005. Available online: https://dokumen.tips/documents/modern-physics-by-serway-moses-and-moyer-third-edition.html (accessed on 20 September 2022).

42. Krane, K.S. *Modern Physics*; John Wiley & Sons, Inc.: Danvers, MA, USA, 2012. Available online: https://scholarsbasics.files.wordpress.com/2018/03/modern-physics-3rd-edgnv64.pdf (accessed on 20 September 2022).
43. Tipler, P.A.; Llewellyn, R.A. *Modern Physics*; W.H. Freeman and Company: New York, NY, USA, 2008. Available online: https://web.pdx.edu/~{}pmoeck/books/Tipler_Llewellyn.pdf (accessed on 20 September 2022).
44. Thornton, S.T.; Rex, A. *Modern Physics for Scientists and Engineers*; Brooks/Cole, Cengage Learning: Boston, MA, USA, 2013. Available online: https://sr.1lib.limited/book/1292592/7401fc (accessed on 20 September 2022).
45. Nolan, P.J. *Fundamentals of Modern Physics*; State University of New York-Farmingdale: Farmingdale, NY, USA, 2014. Available online: https://tekartlearning.com/Nolan_Modern_Physics.pdf (accessed on 20 September 2022).
46. Noce, C. (Ed.) *Modern Physics*; A Critical Approach; IOP Publishing: Bristol, UK, 2020. [CrossRef]
47. Halliday, D.; Resnick, R.; Walker, J.; Taylor, E.F. *Understanding Modern Physics*; Magnes Press: Jerusalem, Israel, 2012. (In Hebrew)
48. Stebbins, C.M.; Coolidge, M.H. *Golden Treasury Readers: Primer*; American Book Co.: New York, NY, USA, 1909.
49. Einstein, A.; Podolsky, B.; Rosen, N. Can quantum-mechanical description of physical reality be considered complete? *Phys. Rev.* **1935**, *47*, 777. [CrossRef]
50. Aspect, A.; Dalibard, J.; Roger, G. Experimental test of Bell's inequalities using time-varying analyzers. *Phys. Rev. Lett.* **1982**, *49*, 1804. [CrossRef]
51. Cheong, Y.W.; Song, J. Different levels of the meaning of wave-particle duality and suspensive perspective on the interpretation of quantum theory. *Sci. Educ.* **2014**, *23*, 1011–1030. [CrossRef]
52. Lévy-Leblond, J.-M. On the nature of quantons. *Sci. Educ.* **2003**, *12*, 495–502. [CrossRef]
53. Krijtenburg-Lewerisa, K. *Teaching Quantum Mechanics at the Secondary School Level*; University of Twente: Enschede, The Netherlands, 2020. [CrossRef]
54. Krijtenburg-Lewerissa, K.; Pol, H.J.; Brinkman, A.; van Joolingen, W.R. Key topics for quantum mechanics at secondary schools: A Delphi study into expert opinions. *Int. J. Sci. Educ.* **2019**, *41*, 349–366. [CrossRef]
55. Greinert, F.; Müller, R. *Competence Framework for Quantum Technologies: Methodology and Version History*; European Commission: Brussels, Belgium, 2021. [CrossRef]
56. Merzel, A.; Weissman, E.Y.; Katz, N.; Galili, I. Toward teachers training for teaching quantum physics in high school. In *Physics Teacher Education–What Matters?* Marks, J.B., Galea, P., Gatt, S., Sands, D., Eds.; Springer Nature Switzerland: Cham, Switzerland, 2022; pp. 161–172. [CrossRef]
57. Galili, I. Towards a refined depiction of nature of science: Applications to physics education. *Sci. Educ.* **2019**, *28*, 503–537. [CrossRef]
58. Weissman, E.Y.; Merzel, A.; Katz, N.; Galili, I. Teaching quantum physics as a structured physics theory in high school. *J. Phys. Conf. Ser.* **2021**, *1929*, 012051. [CrossRef]
59. Linn, M.C. Designing the Knowledge Integration Environment. *Int. J. Sci. Educ.* **2000**, *22*, 781–796. [CrossRef]
60. Xu, W.; Liu, Q.; Koenig, K.; Fritchman, J.; Han, J.; Pan, S.; Bao, L. Assessment of knowledge integration in student learning of momentum. *Phys. Rev. Phys. Educ. Res.* **2020**, *16*, 010130. [CrossRef]

physics

MDPI

Article

Introducing Quantum Technologies at Secondary School Level: Challenges and Potential Impact of an Online Extracurricular Course

Maria Bondani [1,*], Maria Luisa Chiofalo [2], Elisa Ercolessi [3], Chiara Macchiavello [4], Massimiliano Malgieri [4], Marisa Michelini [5], Oxana Mishina [6], Pasquale Onorato [7], Filippo Pallotta [8], Sara Satanassi [3], Alberto Stefanel [5], Claudio Sutrini [4], Italo Testa [9] and Giacomo Zuccarini [4]

1 Istituto di Fotonica e Nanotecnologie-CNR-IFN, 22100 Como, Italy
2 Dipartimento di Fisica, Università di Pisa, 56122 Pisa, Italy
3 Dipartimento di Fisica e Astronomia "Augusto Righi", Alma Mater Studiorum Università di Bologna, 40126 Bologna, Italy
4 Dipartimento di Fisica, Università di Pavia, 27100 Pavia, Italy
5 Unità di Ricerca in Didattica della Fisica, Università di Udine, 33100 Udine, Italy
6 Istituto Nazionale di Ottica-CNR-INO, c/o SISSA–Scuola Internazionale Superiore di Studi Avanzati, 34136 Trieste, Italy
7 Dipartimento di Fisica, Università di Trento, 38123 Trento, Italy
8 Dipartimento di Scienza e Alta Tecnologia-Università dell'Insubria, 20100 Como, Italy
9 Dipartimento di Fisica "E. Pancini", Università Federico II, 80126 Napoli, Italy
* Correspondence: maria.bondani@uninsubria.it

Abstract: Stimulated by the European project "QTEdu CSA", within the flagship "Quantum Technologies", a community of researchers active in the fields of quantum technologies and physics education has designed and implemented an extracurricular course on quantum physics concepts and quantum technologies applications for high school. The course, which featured eight interactive lectures, was organized online between March and May 2021 and attended by about 250 students from all over Italy. In this paper, we describe the main tenets and activities of the course. Moreover, we report on the effectiveness of the course on students' knowledge of the basic concepts of quantum physics and students' views about epistemic aspects and applications of quantum technologies. Results show that the designed activities were effective in improving students' knowledge about fundamental aspects of quantum mechanics and familiarizing them with quantum technology applications.

Keywords: quantum physics; structural equation modeling; secondary school students

1. Introduction

Industrial policies in the EU and US in the past six years have fostered a fast and increasing interest in quantum mechanics as a pivotal area for the development of future technological and societal advancements [1]. In particular, quantum mechanics is at the heart of innovations that include intelligent sensors, networking, communication, computing hardware, algorithms, and other facilitating technologies. Such interest has led to the launch of far-reaching institutional programs such as the *National Quantum Initiative Act* [2,3] in the US and the *Quantum Flagship* in the EU [4]. However, this interest vividly clashes with physics education research evidence according to which quantum mechanics is perceived as a difficult and demanding subject area, whose concepts are considered too abstract and difficult [5–10]. Many studies also show that students hold a variety of misconceptions in quantum mechanics [11–13] due, for instance, about the need to reconsider the key concepts of classical physics and to bridge physics and chemistry concepts [14–16]. In this study, we propose to use quantum technologies as a suitable educational context to introduce foundational aspects of quantum mechanics at the secondary school level.

Our proposal also builds on recent calls by stakeholders in the EU and US to meet future demand for a well-trained and aptly prepared workforce to be employed in this promising industry field [1].

Stimulated by the European project "QTEdu CSA", active within the *Quantum Flagship* program we have designed and implemented an educational path for high school extracurricular activities on "Quantum Technologies" aimed at introducing basic concepts of quantum physics in an effective way taking advantage of the context of the second quantum revolution [17].

The activity is the result of the joint efforts of the Italian communities of researchers active in the fields of quantum technologies and physics education.

1.1. *The Use of Quantum Technologies to Introduce Quantum Mechanics*

We decided to use the context of quantum technologies to convey the concepts of quantum mechanics. Our basic assumption is that quantum technologies may reduce the students' perceived abstractness of quantum mechanics, which often comes from limited access to suitable experimental and mathematical literacies.

The basic idea of the educational path is to describe the logic of quantum physics by establishing a parallelism with the logic circuits of information theory [18–20]. The axioms of quantum mechanics describe the preparation of a state, its evolution/manipulation, and its measurement, which can be interpreted as information input, information processing, and information output, respectively. This parallelism makes it possible to introduce the fundamental properties of quantum states (superposition and entanglement) and to introduce the "qubit," the quantum extension of the classical "bit," and the elementary transformation of the qubit in terms of quantum gates. Simulations and descriptions of experiments with spin and polarization are used to discuss the physical implementation of qubits. The radical novelty introduced by quantum theory becomes clear from the analysis of the superposition state, the meaning of probability, and the role of measurement.

1.2. *The Educational Path to Quantum Technologies*

The activities of our learning path were structured in the following steps: four introductory lectures (one hour each), an in-depth course of three lectures (one and a half hours each) on specific aspects of quantum technologies, and a closing lecture (one and a half hour); see Figure 1.

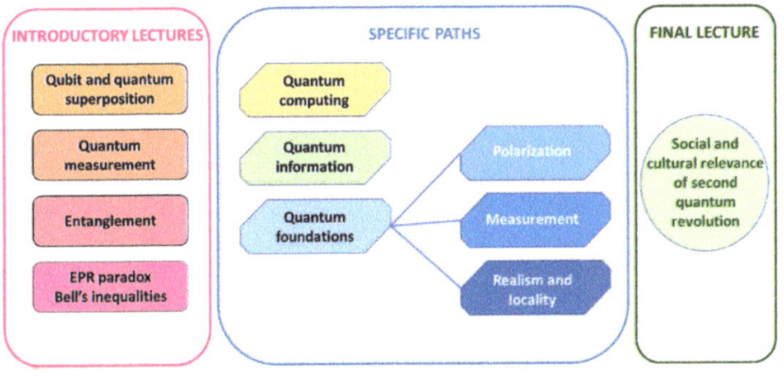

Figure 1. Overview of the educational path activities.

Going into more detail, in the first lecture, starting with the example of the classical and quantum coin flip, we introduce the concepts of quantum superposition state (qubit) and the possible transformations on a single qubit described as logic gates (Figure 2).

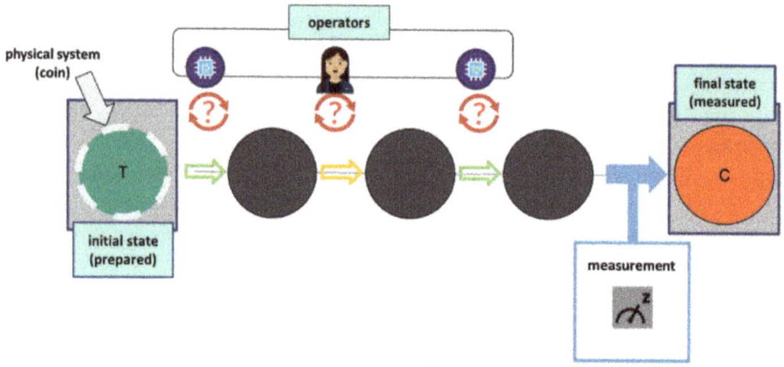

Figure 2. First lecture: the quantum coin flip. Schematics of the quantum coin flip "thought" game interpreted in terms of state preparation, evolution and measurement. One coin is initially prepared in a given state (T, i.e., Head here). After a number of flips by either the quantum computer (displayed as a chip) or the human operator, a measurement is performed (here, yielding C, i.e., Tail).

In the second lecture, we introduce the use of the IBM Q-composer [21] to write and execute quantum circuits on simulators and real quantum computers (Figure 3).

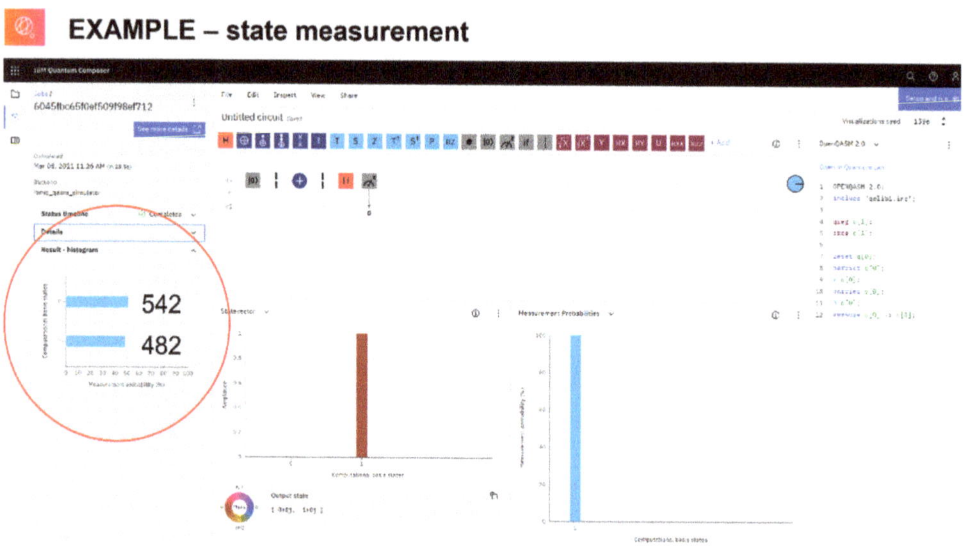

Figure 3. Second lecture: IBM Q-composer interface for programming quantum computers. Example of the quantum coin flip circuit executed on a simulator. In the Q-composer interface, $|0\rangle$ and $|1\rangle$ are the kets representing the computational basis. Center of the graphic interface. **Top**: the row of symbols represents a number of quantum logic operators (quantum gates) that can be picked up to compose the quantum circuit. **Middle**: below the list of operators is the composed circuit: here, to the $|0\rangle$ state an X-not gate (represented by the plus symbol) is applied, followed by an Hadamard gate (H symbol) and a measurement operation (gauge symbol). **Bottom**: representation of the state vector in terms of probability and in the Bloch sphere representation (**left**) and of the measurement probabilities (**right**). A single measurement yields just one of the possible outcomes (1 in this case). The left part of the graphical interface shows an output of a repeated measurement yielding a statistical distribution of the two possible outcomes (in the red circle). The right part of the interface shows Python code corresponding to the quantum circuit. Credits: IBM Q-composer [21].

Quantum mechanical applications require measuring the state of the qubits; thus we introduce the rules of quantum measurement, its epistemological aspects, and its role in the implementation of quantum algorithms. The lecture is supported by the QuVis simulator for a series of Stern–Gerlach apparatuses on spin-1/2 particles (Figure 4) [22].

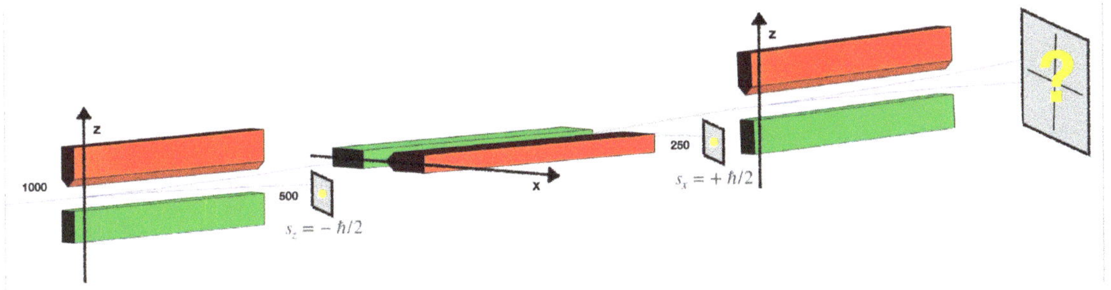

Figure 4. Third lecture: QuVis simulation of Stern–Gerlach experiment. 1000 spins pass through a z-magnet, after which a measurement is performed detecting 500 of the spins in a spin-down state (along z). The remaining spin-up states pass through an x-magnet, after which a measurement is performed detecting half of the spins in a spin-up state (along x). The remaining 250 spins finally pass through a second z-magnet. The students are then asked (notified by a question mark) what they do expect would be the outcome of a measurement, after the latter process. $s_z = -\hbar/2$ and $s_z = +\hbar/2$ define the spin down and up z-components, respectively, and $\hbar$ is the Planck's reduced constant.

Particular care was paid to describe the role of non-commuting quantum operation and the need to describe the spin state of the particle as a superposition state.

In the third lecture, we address two-qubit states and gates to introduce the concept of entanglement, both from the physical and the algorithmic point of view. The IBM Q-interface is used for the Bell-state circuit.

The relevance of entangled states for quantum technologies is highlighted through the description of the BBM92 protocol for secure cryptographic key generation (Figure 5) and the Schrödinger cat paradox.

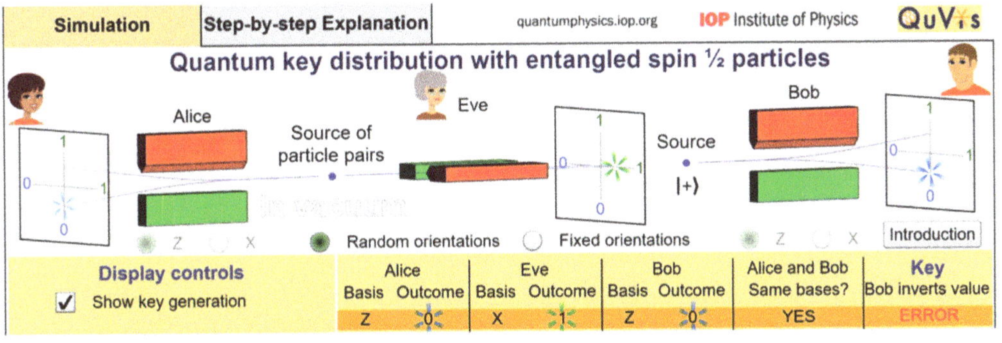

Figure 5. Third lecture: QuVis simulation of BBM92 cryptographic protocol in the presence of an eavesdropper. Alice and Bob share a source of entangled spin-1/2 particles and measure through Z-magnet or X-magnet to obtain a secret key. The eavesdropper Eve measures the spin state of the transiting qubit toward Bob. Eve cannot measure simultaneously along Z and X and therefore cannot obtain complete information about the spin state of the intercepted qubit. When Eve resends the qubit to Bob the spin can be different from the original one; for details, see [23].

The fourth lecture is devoted to a general recap of the concepts introduced in the first three lectures to stress the overall rationale and the links among them. A part of the final lecture introduces the Einstein-Podolsky-Rosen (EPR) paradox and the violation of Bell's inequalities (Figure 6) [24].

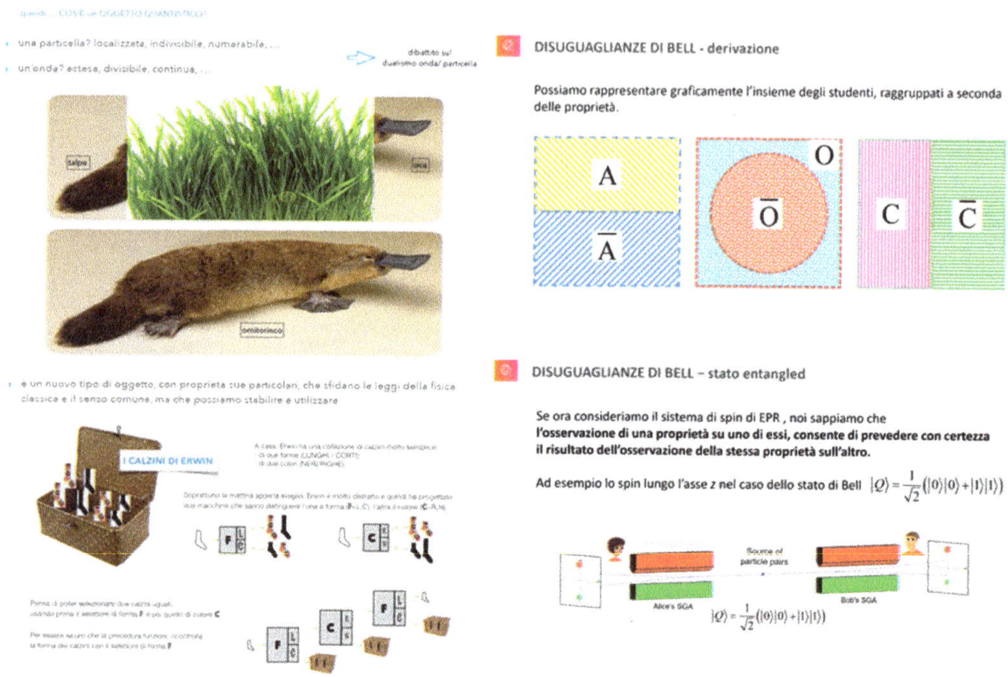

Figure 6. Fourth lecture: recap of the concepts and discussion of theory interpretations. Screenshots of the slides used during the lecture. **Left**: What is a quantum object? **Right**: concept of properties (**top**) and entanglement (**bottom**).

The second phase of our learning path is devoted to specialized lectures in which quantum technologies are examined from both physical and technological points of view. Three different paths were designed: quantum computing, quantum communications, and quantum mechanics foundations, with the latter organized into the three sub-paths: polarization; time evolution and the measurement paradox; realism and locality.

Finally, in the last phase, the state of art quantum technologies is from the viewpoint of their social and economic impact by examining the appearance of the targeted concepts in the media.

More details on the educational path are included in Appendix A.

1.3. Aims of the Study

This study has a twofold aim: (1) building on the theoretical background described above, to evaluate whether the designed path helped the students to grasp a basic knowledge of fundamental quantum physics; (2) building on the potential attractiveness of the quantum technologies applications, to evaluate whether the designed didactical path improved students' views about quantum technologies. Thus, we posit the following research questions:

To what extent was the educational path on quantum technologies effective in improving the students' knowledge about fundamental quantum mechanics concepts?

To what extent was the educational path on quantum technologies effective in improving the students' views about quantum technologies?

2. Methods and Tools

2.1. Instructional Context

The study was carried out in the context of the Italian plan called Paths for Transversal Competencies and Orientation (PCTO). The PCTO activities are mandatory for students and include either career orientation or vocational practice.

The PCTO activities on quantum technologies here reported came about as a follow-up to the outreach conference on the second quantum revolution organized by the European project QTEdu in November 2020. About 2000 students have registered for the conference and more than 3800 have viewed it on YouTube so far [25]. The conference aimed at giving a general introduction to the different quantum technologies that emerged from the first quantum revolution and are emerging from the second one, avoiding overly technical and explicit physical and mathematical content. Building on the success of the conference, we decided to offer Italian high-school teachers and students the opportunity to deepen their knowledge on the topic.

The activities described in this study were carried out in spring 2021 for a total time of about 12 h over a time span of two months, in remote distance modality using the Zoom platform. Lectures were conducted by alternating presentations with simulations and exercises. Student engagement was stimulated by interactive questions and clickers to maintain attention and check their understanding of the content.

2.2. Sample

The sample consisted of N = 279 Italian high-school students (females: 24.4%; males: 73.8%; prefer not to say: 1.8%) from 16 different schools distributed across Italy. The course was restricted to students attending the 12th (N = 101, average age: 18.0 ± 0.4 s.d. (standard deviation)) and 13th grades (N = 178; average age: 19.0 ± 0.5 s.d.). The great majority (82%) of the students attended the Liceo Scientifico (mathematically-oriented high school), about 12% attended an applied science course (natural sciences-oriented high school), and about 6% attended a technical school.

2.3. Instruments

2.3.1. Quantum Technologies Inventory (QTI)

For the present study, we developed a short questionnaire (QTI) featuring eight items on the topics addressed during the educational path. The questionnaire was built on our prior studies [26–29]. The reason for developing such a new instrument is that none of the current instruments for evaluating students' performance in quantum mechanics focuses on the central concepts addressed in our educational path about quantum technologies. Table 1 summarizes the concepts and topics addressed in the questionnaire. The complete questionnaire is reported in Appendix B.

Table 1. The main topics addressed in the quantum technologies inventory (QTI).

Concept	Sub-Topic	Item
State	Bra-Ket Formalism	Q3 Q4 Q7
	State and eigenstate in quantum physics	Q1 Q2 Q3
	Logical gate	Q1
	Qubit	Q3
Superposition		Q1 Q7
Measurement	Statistical nature of the measurement	Q2 Q6 Q8
	Wave function collapse	Q7
Entanglement	Formalism	Q4
	measurement on entangled states	Q5

2.3.2. Views about Quantum Technologies (VAQT)

Five elements were used to measure students' views about quantum, technologies:

1. Assuming to use ideal measuring instruments, in physics I must describe the results of measurements probabilistically only if I have incomplete information about the system;
2. It is possible for physicists to carefully perform the same experiment and get two very different results that are both correct;
3. Quantum computers will never work, because it is impossible to build an hardware that is accurate enough;
4. Scientists say that quantum communication makes it possible to teleport a particle from one place to another;
5. Scientists say that quantum communication does not make possible the teleportation of a particle from one place to another, but only the transfer of its characteristics.

The elements 1 and 2 address students' general epistemic views about quantum mechanics, namely ideas about values, assumptions, processes, and formations of knowledge about quantum mechanics [30]. Elements 3–5 address more specific views about quantum technologies applications.

2.4. Data Analysis

The QTI items were scored as follows. Correct answers to two items (Q1, Q3) received 2 points, while one option received 1 point. For item Q8, two answer choices received 1 point. Correct answers to the remaining six items received 1 point, while the remaining options received no credit. The reason for such scoring was that Q1 and Q3 addressed more complex topics that allowed for a more refined scoring, while Q8 had two correct answer choices. The total score for the eight items was therefore 10. The five VAQT items were scored using a 5-point Likert scale.

The QTI was submitted after the teaching activities. The VAQT was submitted before and after the teaching activities. Pre-post differences in the VAQT items were evaluated through a *t*-student test and Cohen's effect size, *d* [31].

3. Results

Overall, 176 students responded to the QTI after the didactical path, while 162 completed the VAQT instrument before and after the out-of-school activities.

The breakdown of the students' answers to the eight items of the QTI instrument is given in Table 2. The distribution of the students' scores is s hown is in Figure 7.

Table 2. Frequencies of N = 176 students' responses to the QTI. Correct answer choices are in boldface. See Appendix B for the complete questionnaire.

Answer Choice	Q1	Q2	Q3	Q4	Q5	Q6	Q7	Q8
A	**20%**	7%	52%	17%	14%	**31%**	**38%**	9%
B	36%	30%	**15%**	**40%**	18%	10%	22%	5%
C	39%	**53%**	18%	16%	15%	23%	11%	**32%**
D	3%	9%	13%	22%	**51%**	33%	27%	**55%**
No answer	1%	1%	2%	4%	3%	3%	2%	1%

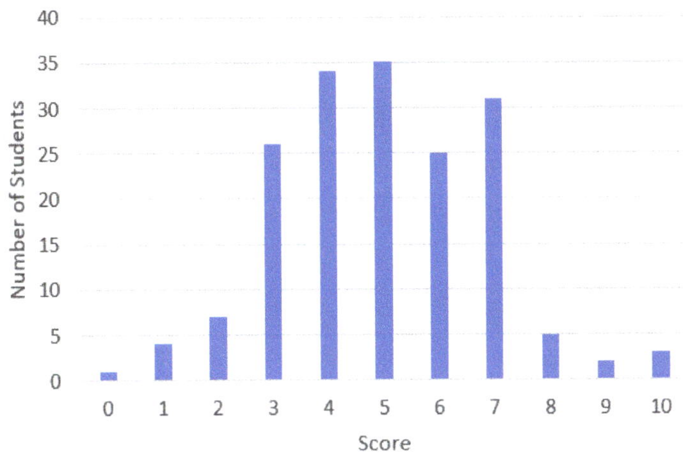

Figure 7. Distribution of students' scores in the quantum technologies inventory (QTI). See text for details.

The average pre-instruction score in the VAQT items about general epistemic aspects was 3.04 ±.85 s.d., while the average post-instruction score was 3.40 ± 0.91 s.d. The difference is statistically significant (*t*-test $t = 4.68$, degrees of freedom df = 160, probability $p < 0.001$), with a medium effect size (Cohen's $d = 0.37$) [31]. The average pre-instruction score in the VAQT items about quantum technologies applications was 3.43 ± 0.74 s.d., while the average post-instruction score was 3.65 ± 0.80 s.d. The difference is statistically significant ($t = 3.35$, df = 160, $p < 0.001$), with a medium effect size (Cohen's $d = 0.26$).

To better understand this trend, we divided the sample into four groups according to their performance and then calculated the effect size of the difference between the post and pre-test for the VAQT items for each group. Figures 8 and 9 show the average pre-instruction and post-instruction scores for each group, while Table 3 gives the corresponding Cohen's *d*.

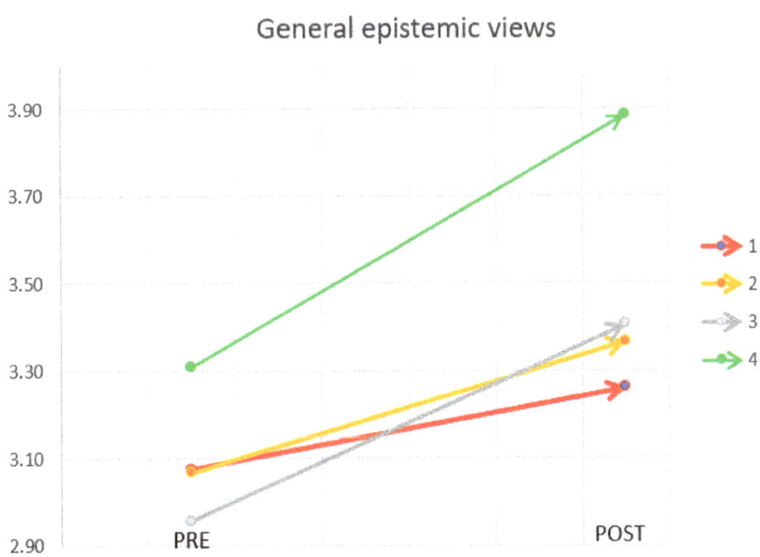

Figure 8. Pre-instruction vs. post-instruction average scores of the views about quantum technologies (VAQT) instrument (general epistemic views) according to the performance in the QTI for four groups (see Table 3).

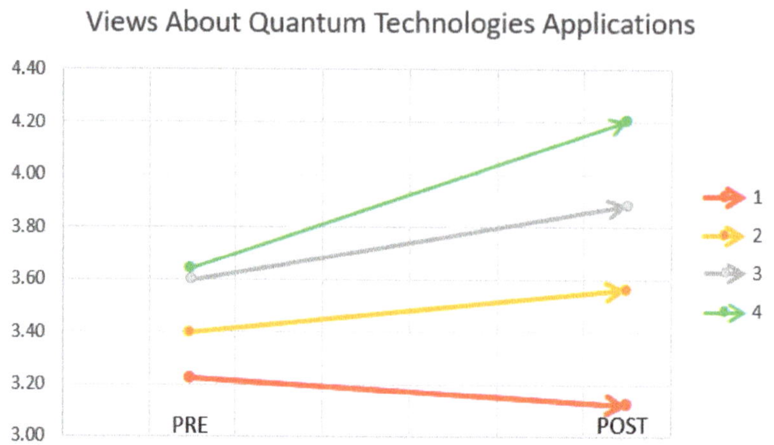

Figure 9. Pre-instruction vs. post-instruction average scores of the VAQT instrument (applications of quantum technologies) according to the performance in the QTI for four groups (see Table 3).

Table 3. Sample subdivision and pre-post effect size for the views about quantum technologies (VAQT) instrument according to QT performance score.

Group	QTI Score	Number of Students (N = 162)	Cohen's *d*	
			General epistemic views	QT applications
G1	0–3	25	0.23	−0.14
G2	4–5	71	0.36	0.22
G3	6–7	56	0.49	0.38
G4	8–10	10	0.54	0.96

4. Discussion and Conclusions

The analysis of the students' answers to the QTI shows that, on average, the educational path was useful to familiarize students with fundamental aspects of quantum mechanics. Considering the fact that this was the first time the students tackled central concepts of quantum theory, the educational path was effective in focusing students' attention on the fundamental concepts of quantum mechanics through the parallelism of information theory. About 40% of the students show a good knowledge of the measurement concept and of the wave function collapse, as measured by the QTI items. Similarly, on average, about 45% of the students seem to have grasped the concept of entanglement, which is not included in the national curriculum. Only about one-third of the students correctly answered questions about the state and superposition. However, the students found it difficult to apply formal calculations to states represented as qubits. On such a basis, we are currently revising our educational path to improve the presentation of the concepts of state, qubit, and superposition. We also plan to improve the instruments used to assess the effectiveness of the proposed activities.

From the analysis of the students' responses to the VAQT, it emerges that, on average, the proposed path was effective in letting students acquire more informed views about general epistemic aspects of quantum mechanics and quantum technology applications. Greater gains, as measured by the Cohen's *d* effect size, were shown by the students who performed better on questions about the fundamental concepts of quantum mechanics. Such a result suggests a link between the knowledge of quantum physics concepts and informed views about epistemic aspects. However, more research is warranted to investigate such a relationship in more detail.

Limitations of the study include the involvement of a small sample and the remote distance modality of the didactical activities. Results are also limited by the use of instruments not yet validated. Oral interviews would have improved the reliability of our findings. We plan to revise both instruments and perform a more rigorous study of their validity and reliability.

Overall, the present study shows that the proposed educational path may be effective in helping students grasp fundamental quantum concepts, acquire informed views about how knowledge in quantum mechanics is constructed, and familiarize them with applications of quantum technologies.

The first implementation of our didactical path is encouraging and may help other researchers interested in quantum mechanics education to implement similar activities. While promising, our results also point out that initial motivation toward physics is crucial to grasping the addressed concepts. This evidence calls for a stricter collaboration with schoolteachers with the aim to share the key ideas and didactical objectives of the activities. With this in mind, a second run of PCTO activities was performed in spring 2022 with a very similar program but with fewer students attending. Data analysis is ongoing. Moreover, the same approach to quantum physics, more focused on basic concepts, presentation of axioms, and quantum technologies was adopted by the outreach project "Italian Quantum Weeks" (2022–2024) [32] devoted to students and the general public.

Author Contributions: Conceptualization, M.B., M.M. (Massimiliano Malgieri), P.O. and I.T.; methodology, M.B., M.M., P.O. and I.T.; software, M.B., F.P., S.S. and G.Z.; validation, M.L.C., E.E., C.M., O.M., A.S., C.S. and G.Z.; formal analysis, P.O. and I.T.; investigation, M.B., M.M. (Massimiliano Malgieri) and I.T.; resources, O.M.; data curation, P.O. and I.T.; writing—original draft preparation, M.B. and I.T.; writing—review and editing, M.L.C., E.E., C.M., M.M. (Massimiliano Malgieri), M.M. (Marisa Michelini), O.M., P.O., F.P., S.S., A.S., C.S., I.T. and G.Z.; visualization, M.B.; supervision, M.B., M.M., P.O. and I.T.; project administration, M.B.; funding acquisition, M.B. All authors have read and agreed to the published version of the manuscript.

Funding: This research received no external funding.

Data Availability Statement: The data presented in this study are available on request from the corresponding author. The data are not publicly available due to privacy issues.

Conflicts of Interest: The authors declare no conflict of interest.

Appendix A. Path Description

Appendix A.1. Plenary Introductory Lectures

- **Lecture 1** Objective: quantum state, superposition state Content: qubits, Dirac representation, role of probability in quantum mechanics
    1. Classical and quantum coin-flipping game
    2. How to describe a quantum coin: bits and qubits
        - Dirac notation
        - Superposition states
        - Visual representation of qubits (Bloch sphere)
    3. Operations on quantum coin: single qubit quantum logic gates
        - I, NOT, Z, H
    4. Physical examples of qubits
        - Single photon after a beam splitter (QuVis, MILQ [33])
        - Electronic spin
        - Cavity atom
        - Superconducting circuits
- **Lecture 2** Objective: measurement Content: representation of quantum measures, role of probability, non-compatible measurements
    1. Introduction to IBM Quantum experience

    - again on the logic of H, NOT, Z, X, and exercises on IBMQ
    - definition of coefficients of superposition states (probability amplitude), probability, and measurement
  2. Stern–Gerlach experiment
    - simulated experiments for sequences of Stern-Gerlach apparatuses performed using QuVis
    - discussion on the logical necessity of describing the spin state of the electron as a superposition state
  3. Experiments with QuVis
    - difference between mixtures and superposition states
    - sequence of three Stern–Gerlach apparatuses
- **Lecture 3** Objective: entanglement Content: separable and entangled states, correlated measurements
  1. Introduction
    - two-qubit states (basis elements, normalization)
    - difference between entangled state and separable state
    - measurement of single qubits in many-qubit states
    - Bell states: invariance upon basis rotation
    - reprise of Stern–Gerlach experiments to discuss the value of reality to be attributed to the variables
  2. IBMQ interlude
    - quantum circuit to generate entangled states
  3. Application of entanglement
    - QuVis: cryptographic protocol BBM92
  4. Historical contextualization
    - Einstein-Podolsky-Rosen (EPR) paradox (Schrödinger's cat): locality and realism
    - Bell's inequalities: classical correlations vs entanglement
    - QuVis: cryptographic protocol BBM92
- **Lecture 4** Objective: synthesis of concepts and formalism Content: axiomatic conceptual framework of quantum mechanics, interpretative problems
  1. Summary of the concepts discussed in the previous lectures
    - highlight the connections among the concepts
    - introduce concepts intentionally not mentioned in the lectures but that the students will encounter in their course of study: uncertainty principle, wave function, dualism
  2. Interpretative problems in quantum mechanics
    - EPR paradox
    - Bell's inequalities

Appendix A.2. Specific Paths

- **Quantum computing** (Figure A1)

  Objective: basics of quantum computing
  Content: quantum gates, quantum algorithms, teleportation, quantum random walk

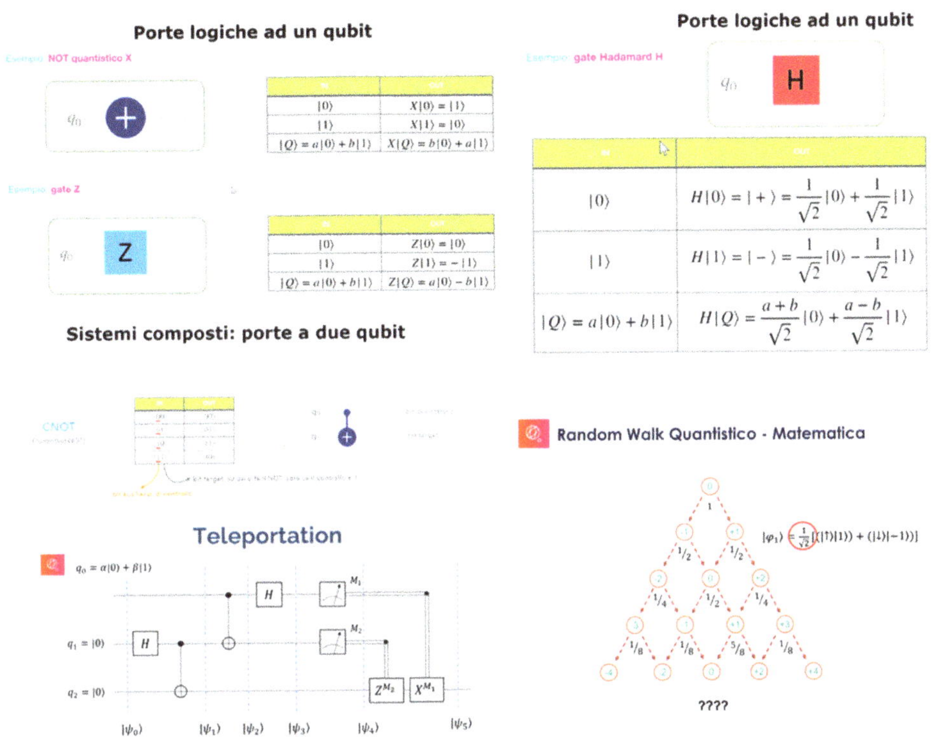

Figure A1. Examples from the specific path "Quantum computing". Screenshots of the slides used during the lecture. **Top**: single-qubit gates. **Left middle**: two-qubit gates. **Left bottom**: teleportation protocol. **Right-bottom**: one step of the mathematical description of the quantum random walk protocol.

- **Quantum information** (Figure A2)

  Objective: basics of quantum information

  Content: classical and quantum cryptography, cryptographic protocols, quantum random number generation for cryptography

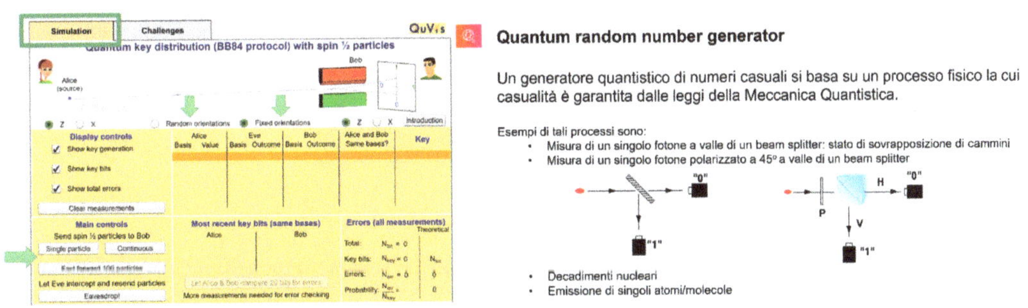

Figure A2. Examples from the specific path "Quantum information". **Left**: BB84 protocol implemented on QuVis. **Right**: example of implementation of a random number generator with single photons.

- **Fundamentals: Polarization** (Figure A3)

  Objective: understand basic principles

  Content: state, properties, superposition principle, entanglement

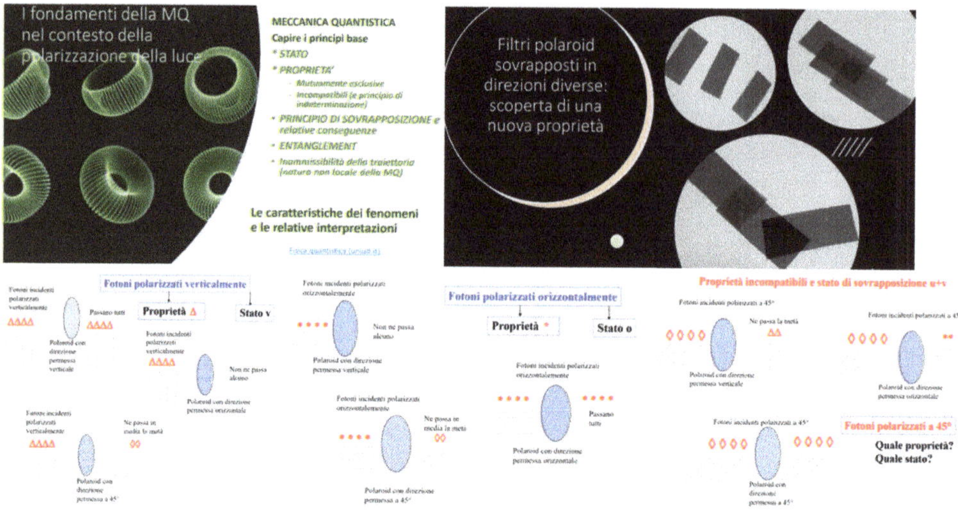

Figure A3. Examples from the specific path "Fundamentals: Polarization". Screenshots of the slides used during the lectures. **Top left**: summary of the lectures. **Top right**: examples of polaroid superposition. **Bottom**: schematics of the behavior of polarized photons passing through polaroid filters.

- **Fundamentals: Measurement Paradox** (Figure A4)

  Objective: evolution of quantum systems and measurement
  Content: the laws of quantum mechanics, logic gates, Schrödinger's cat

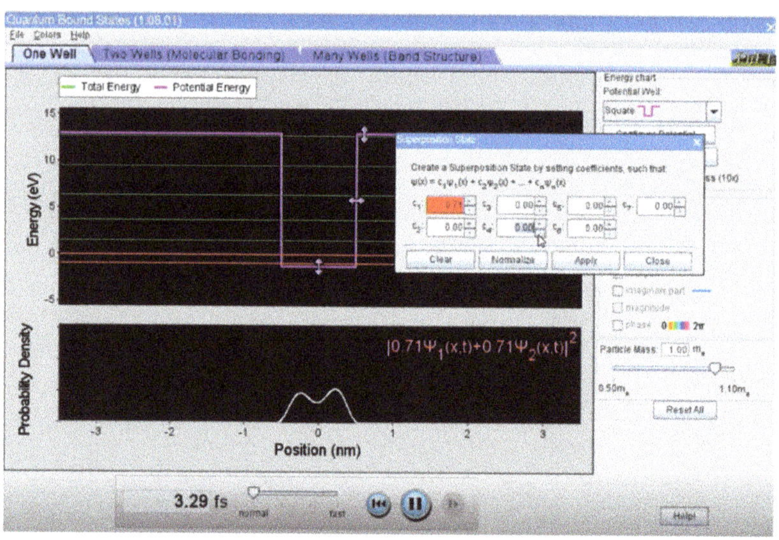

Figure A4. Examples from the specific path "The measurement paradox: Schrödinger's cat and other fantastic beasts (and where to find them)". Before comparing system evolution governed by the Schrödinger equation and evolution in measurement, students are led to use a PhET simulation in order to elicit qualitative properties of the former.

- **Fundamentals: Realism and Locality** (Figure A5)

  Objective: discussion of interpretative problems in quantum mechanics

Content: EPR paradox, hidden variable theories, Bell's inequalities

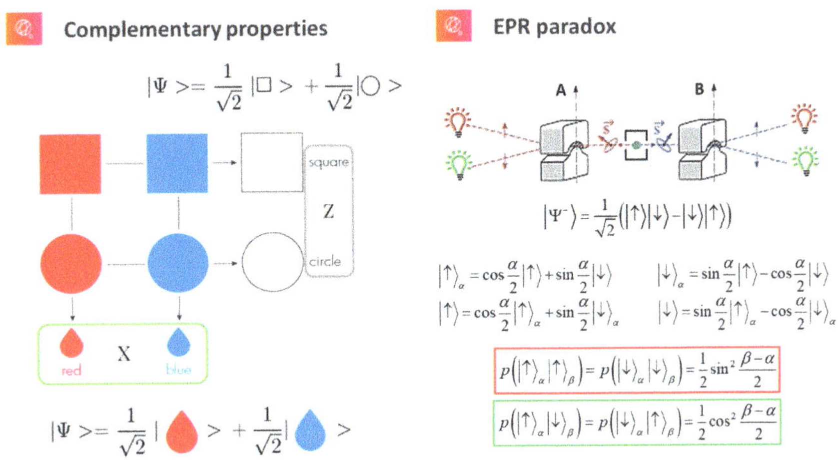

Figure A5. Examples from the specific path "Fundamentals: Realism and locality". Left panel: pictorial representation of complementary properties. Right panel: scheme of the EPR paradox experiment and mathematical description of the entangled state and the values of detection probabilities for complementary measurements.

Appendix A.3. Final Lecture (Figure A6)

Objective: social and cultural relevance of the second quantum revolution

Content: presentation of the current debate and the implications of the second quantum revolution at national and international levels through newspaper articles and official documents (e.g., Quantum Manifesto) description of present quantum technologies, description of new jobs in quantum technologies.

Figure A6. Examples from the final lecture. Screenshots of the slides used during the lecture. Top left: timeline of quantum computer technology. Top right and bottom-left: examples of newspaper titles on quantum computers. Bottom-right: image from the website of the *Quantum Technology Flagship*.

Appendix B. Quantum-Technologies (QT) Questionnaire

We report in boldface the correct answers (see Table 2). The percentage for each answer choice is also reported. Note: "NR" denotes "not responded".

Q1. A quantum state prepared in a $|0\rangle$ state undergoes a transformation (logic gate, beam splitter, magnetic field acting on the spin) but no measurement is made on the system. Will the resulting state after the transformation (final state) be uniquely determined? (NR = 1%):

(a) Yes. Knowing how the transformation changes the initial state, the final state is uniquely determined (20%)
(b) No. You could get different states because the transformation might not end up on a well-defined state $|0\rangle$ or $|1\rangle$ but on an intermediate state (36%)
(c) No, because I can obtain states with different probability defined by the superposition coefficients (39%)
(d) Yes, the state remains the same regardless the transformation (3%)

Q2. To "know" the state of a quantum system implies that we are able to predict: (NR = 1%):

(a) the outcome of any measure on the system with certainty (7%)
(b) the outcome of a single measurement of any observable within the limits set by the uncertainty principle (30%)
(c) the probability of different outcomes of any given measure on the system (53%)
(d) the state a system after any measurement process with certainty (9%)

Q3. A single qubit is on a superposition state $|Q\rangle = a|0\rangle + b|1\rangle$ and no measurement is performed. That means (NR = 2%):

(a) The state of the single qubit $|Q\rangle$ is $|0\rangle$ with probability $p0 = |a|^2$ or $|1\rangle$ with probability $p1 = |b|^2$ (52%)
(b) The state of the single qubit $|Q\rangle$ is different from both $|0\rangle$ and $|1\rangle$ (15%)
(c) The state of the single qubit $|Q\rangle$ is both $|0\rangle$ and $|1\rangle$ (18%)
(d) The state of the single qubit $|Q\rangle$ oscillates between $|0\rangle$ and $|1\rangle$ (13%)

Q4. Considering these two-qubit states $|\Psi1\rangle = 1/\sqrt{2}(|0\rangle|1\rangle + |1\rangle|0\rangle)$ and $|\Psi2\rangle = 1/\sqrt{2}(|0\rangle|0\rangle - |1\rangle|0\rangle)$. We can then say that (NR = 4%):

(a) $|\Psi1\rangle$ is separable (can be factorized), $|\Psi2\rangle$ is entangled (17%)
(b) $|\Psi2\rangle$ is separable (can be factorized), $|\Psi1\rangle$ is entangled (40%)
(c) Both states are separable (can be factorized) (16%)
(d) Both states are entangled states (22%)

Q5. Consider a system made of two parts A and B (e.g., two electrons and two photons), that is in an entangled state (e.g., of spin or polarization). Having performed a spin or polarization measurement on A and noted the result, we can deduce that (NR = 3%):

(a) The same measurement made on system B gives the same result (14%)
(b) The same measurement made on system B gives the opposite result (19%)
(c) We will be able to predict with certainty the outcome of any measurement made on system B (15%)
(d) We will be able to predict with certainty the outcome of the same measurement made on system B (51%)

Q6. Consider the following two probabilistic predictions. Prediction 1) In the classic coin toss, the outcome "head" has probability 1/2 to occur. Prediction 2) The interaction of a quantum system with a measurement device can only result in two outcomes (e.g., spin $|\downarrow\rangle$ or $|\uparrow\rangle$, vertical or horizontal polarization), each with probability 1/2. Why is it that in both cases the only predictions we are capable of making are probabilistic? (NR = 3%):

(a) In prediction (1), we do not know the initial conditions precisely enough. In prediction (2) even if the initial conditions are known, the outcome of the interaction is inherently uncertain. (31%)
(b) In both predictions, we do not know the initial conditions precisely enough (10%)

(c) In prediction (1), we do not know the initial conditions precisely enough. In prediction (2) even if the initial conditions are known, we do not know how the interaction works precisely enough (23%)
(d) To be able to make non-probabilistic predictions, in both cases we would need to have perfect control of the experiment, which is not currently possible (33%)

Q7. Assume that a spin measurement is made on the electron in the spin state $|\Psi\rangle = 1/\sqrt{2}(|\ \rangle_z + |^{-}\rangle_z)$ along the Z direction is ↑, i.e., $s_z = +h/2$. What can I say about its spin along Z before measurement? (NR = 2%):

(a) it had no defined value (38%)
(b) it was ↑ already, but the experimenter didn't know (22%)
(c) it was zero (11%)
(d) it was ↓ the 50% of the time and ↑ the 50% of the time (27%)

Q8. Regarding the predictability of the outcome of a measurement in quantum mechanics, which of the following statements is correct? More than one correct answer is possible (NR = 0.5%):

(a) the outcome of a measurement is always unpredictable (9%)
(b) the outcome of a measurement is always predictable (5%)
(c) the outcome of a measurement is predictable with certainty when the process of preparation and of measure are related to the same observable of the system (32%)
(d) the outcome of the measurement process is always predictable only in terms of the probability of outcomes and never with certainty (54%)

References

1. Fox, M.F.; Zwickl, B.M.; Lewandowski, H.J. Preparing for the quantum revolution: What is the role of higher education? *Phys. Rev. Phys. Educ. Res.* **2020**, *16*, 020131. [CrossRef]
2. H.R.6227—National Quantum Initiative Act. Available online: https://www.congress.gov/bill/115th-congress/house-bill/6227 (accessed on 1 September 2022).
3. Raymer, M.G.; Monroe, C. The US National Quantum Initiative. *Quant. Sci. Technol.* **2019**, *4*, 020504. [CrossRef]
4. Riedel, M.; Kovacs, M.; Zoller, P.; Mlynek, J.; Calarco, T. Europe's Quantum Flagship Initiative. *Quant. Sci. Technol.* **2019**, *4*, 020501. [CrossRef]
5. Ayene, M.; Kriek, J.; Damtie, B. Wave-particle duality and uncertainty principle: Phenomenographic categories of description of tertiary physics students' depictions. *Phys. Rev. ST-PER* **2011**, *7*, 020113. [CrossRef]
6. Ke, J.-L.; Monk, M.; Duschl, R. Learning introductory quantum physics: Sensori-motor experiences and mental models. *Int. J. Sci. Educ.* **2005**, *27*, 1571–1594. [CrossRef]
7. Johnston, I.D.; Crawford, K.; Fletcher, P.R. Student difficulties in learning quantum mechanics. *Int. J. Sci. Educ.* **1998**, *20*, 427–446. [CrossRef]
8. Manilla, K.; Koponen, I.T.; Niskanen, J.A. Building a picture of students' conceptions of wave- and particle-like properties of quantum entities. *Eur. J. Phys.* **2002**, *23*, 45–53. [CrossRef]
9. Tsaparlis, G.; Papaphotis, G. Quantum-chemical concepts: Are they suitable for secondary students? High-school students' conceptual difficulties and attempts at conceptual change: The case of basic quantum chemical concepts. *Chem. Educ. Res. Pract. Eur.* **2002**, *3*, 129–144. [CrossRef]
10. Tsaparlis, G.; Papaphotis, G. High-school students' conceptual difficulties and attempts at conceptual change: The case of basic quantum chemical concepts. *Int. J. Sci. Educ.* **2009**, *31*, 895–930. [CrossRef]
11. Marshman, E.; Singh, C. Framework for understanding the patterns of student difficulties in quantum mechanics. *Phys. Rev. ST Phys. Educ. Res.* **2015**, *11*, 020119. [CrossRef]
12. Krijtenburg-Lewerissa, K.; Pol, H.J.; Brinkman, A.; van Joolingen, W.R. Insights into teaching quantum mechanics in secondary and lower undergraduate education. *Phys. Rev. Phys. Educ. Res.* **2017**, *13*, 010109. [CrossRef]
13. Singh, C.; Marshman, E. Review of student difficulties in upper level quantum. *Phys. Rev. ST Phys. Educ. Res.* **2015**, *11*, 020117. [CrossRef]
14. Kalkanis, G.; Hadzidaki, P.; Stavrou, D. An instructional model for a radical conceptual change towards quantum mechanics concepts. *Sci. Educ.* **2003**, *87*, 257–280. [CrossRef]
15. Bouchée, T.; de Putter-Smits, L.; Thurlings, M.; Pepin, B. Towards a better understanding of conceptual difficulties in introductory quantum physics courses. *Stud. Sci. Educ.* **2022**, *58*, 183–202. [CrossRef]
16. Stefani, C.; Tsaparlis, G. Students' levels of explanations, models, and misconceptions in basic quantum chemistry: A phenomenographic study. *J. Res. Sci. Teach.* **2009**, *46*, 520–536. [CrossRef]

17. Satanassi, S. The challenge of Quantum Manifesto for science education: Designing a teaching module on quantum computers for secondary school students. *Il Nuovo Cim.* **2020**, *43*, 130. [CrossRef]
18. Mermin, N.D. *Quantum Computer Science—An Introduction*; Cambridge University Press: Cambridge, UK, 2007. [CrossRef]
19. Satanassi, S.; Fantini, P.; Spada, R.; Levrini, O. Quantum Computing for high school: An approach to interdisciplinary in STEM for teaching. *J. Phys. Conf. Ser.* **2019**, *1929*, 012053. [CrossRef]
20. Satanassi, S.; Ercolessi, E.; Levrini, O. Designing and implementing materials on quantum computing for secondary school students: The case of teleportation. *Phys. Rev. Phys. Educ. Res.* **2022**, *18*, 010122. [CrossRef]
21. IBM Quantum. Available online: https://quantum-computing.ibm.com/ (accessed on 1 September 2022).
22. Sakurai, J.-J.; Napolitano, J. *Modern Quantum Mechanics*; Cambridge University Press: Cambridge, UK, 2020. [CrossRef]
23. Uncertainty of Spin Measurement Outcomes. Available online: https://www.st-andrews.ac.uk/physics/quvis/simulations_html5/sims/MeasurementUncertainty/measurement-outcome-uncertainty.html (accessed on 1 September 2020).
24. Bell, J.S. *Speakable and Unspeakable in Quantum Mechanics*; Cambridge University Press: Cambridge, UK, 2004. [CrossRef]
25. La Seconda Rivoluzione Quantistica: Presente e Futuro Delle Tecnologie Quantistiche. Available online: https://www.youtube.com/watch?v=_LoWS0MBTu8&t=448s (accessed on 1 September 2022).
26. Michelini, M.; Zuccarini, G. University students' reasoning on physical information encoded in quantum state at a point in time. In Proceedings of the 2014 Physics Education Research Conference, Minneapolis, MN, USA, 30–31 July 2014; pp. 187–190. [CrossRef]
27. Michelini, M.; Stefanel, A. A path to build basic Quantum Mechanics ideas in the context of light polarization and learning outcomes of secondary students. *J. Phys. Conf. Ser.* **2021**, *1929*, 012052. [CrossRef]
28. Scotti di Uccio, U.; Colantonio, A.; Galano, S.; Marzoli, I.; Trani, F.; Testa, I. Design and validation of a two-tier questionnaire on basic aspects in quantum mechanics. *Phys. Rev. Phys. Educ. Res.* **2019**, *15*, 010137. [CrossRef]
29. Zuccarini, G. Analyzing the structure of basic quantum knowledge for instruction. *Am. J. Phys.* **2020**, *88*, 385–394. [CrossRef]
30. Chiou, G.-L.; Lee, M.-H.; Tsai, C.-C. High school students' approaches to learning physics with relationship to epistemic views on physics and conceptions of learning physics. *Res. Sci. Technol. Ed.* **2013**, *31*, 1–15. [CrossRef]
31. Field, A. *Discovering Stastitics Using IBM SPSS Statistics*; SAGE Publications: London, UK, 2013.
32. Italian Quantum Weeks. Available online: https://www.quantumweeks.it/ (accessed on 1 September 2022).
33. MILQ—Teaching Quantum Physics. Available online: https://www.milq.info/en/ (accessed on 1 September 2022).

physics

MDPI

Article

Key Experiment and Quantum Reasoning

Moritz Waitzmann [1], Kim-Alessandro Weber [2], Susanne Wessnigk [3] and Ruediger Scholz [4,*]

1 Institut fuer die Didaktik der Mathematik und Physik, Leibniz Universitaet Hannover, 30167 Hannover, Germany
2 Institut fuer Quantenoptik, Leibniz Universitaet Hannover, 30167 Hannover, Germany
3 QUEST Leibniz Research School, Leibniz Universitaet Hannover, 30167 Hannover, Germany
4 CRC 1227 Designed Quantum States of Matter, Leibniz Universitaet Hannover, 30167 Hannover, Germany
* Correspondence: r.scholz@iqo.uni-hannover.de; Tel.: +49-511-7622850

Abstract: For around five decades, physicists have been experimenting with single quanta such as single photons. Insofar as the practised ensemble reasoning has become obsolete for the interpretation of these experiments, the non-classical intrinsic probabilistic nature of quantum theory has gained increased importance. One of the most important exclusive features of quantum physics is the undeniable existence of the superposition of states, even for single quantum objects. One known example of this effect is entanglement. In this paper, two classically contradictory phenomena are combined to one single experiment. This experiment incontestably shows that a single photon incident on an optical beam splitter can either be reflected or transmitted. The almost complete absence of coincident clicks of two photodetectors demonstrates that these two output states are incompatible. However, when combining these states using two mirrors, we can observe interference patterns in the counting rate of the single photon detector. The only explanation for this is that the two incompatible output states are prepared and kept simultaneously—a typical consequence of a quantum superposition of states. (Semi-)classical physical concepts fail here, and a full quantum concept is predestined to explain the complementary experimental outcomes for the quantum optical "non-waves" called single photons. In this paper, we intend to demonstrate that a true quantum physical key experiment ("true" in the sense that it cannot be explained by any classical physical concept), when combined with full quantum reasoning (probability, superposition and interference), influences students' readiness to use quantum elements for interpretation.

Keywords: physics education; quantum theory; nature of science; scientific literacy; single photon experiments; quantum reasoning; key experiment

1. Intro: Beyond the Classical Horizon

"Few problems of physics have received more attention in the past than those posed by the dual wave-particle properties of light. The story of the solution of these problems is a familiar one. It has culminated in the development of a remarkably versatile quantum theory of the electromagnetic field. Yet, for reasons which are partly mathematical and partly, perhaps, the accident of history, very little of the insight of quantum electrodynamics has been brought to bear on the problems of optics. The statistical properties of photon beams, for example, have been discussed to date almost exclusively in classical or semi-classical terms. Such discussions may indeed be informative, but they inevitably leave open serious questions of self-consistency, and risk overlooking quantum phenomena which have no classical analogues" [1].

One of the central questions in the context of science education is: how should one overcome the conceptual barriers of quantum physics? Research shows that even major support from the quantum mechanical formalism would help only very little in overcoming these conceptual barriers [2]. Aiming to understand the puzzling quantum phenomena without converting them into quasi classical phenomena with a hybrid status between the quantum and the classical domain is something that requires consistent basic concepts and proper diction. Educational research shows that a quantum physics course focusing on an analysis, which is restricted to a limited number of well-considered phenomena and displays basic and exclusively quantum traits, promotes a deeper conceptual understanding [3].

Some topics from contemporary research on quantum optics are now mandatory course content for advanced quantum physics courses, while coherent optics and experimental approaches are developed for the undergraduate level [4–8]. Here, we discuss the impact of a key experiment combined with the rigorous usage of an appropriate argumentation derived from central quantum traits to obtain access to the quantum domain [8].

Interestingly enough, in and of itself, the specific and extremely successful language for formulating theoretical models in physics is mathematics. At the same time, to be a physical theory, a mindset has to be set up on a natural idea of the experiments, the measurements and the objects of nature to which the mathematical constructs refer. The formalism has to be interpreted upon the background of nature [9]. This vital need usually produces no problems in classical physics because the physical variables and their temporal development (trajectories in phase space) are manifest. Interpretations in quantum physics are much more challenging, because the theoretical constructs (quantum state, probability amplitude, superposition of states, quantum interference) are not just directly and conceptually connected to the real world of nature.

In a recent paper, Bitzenbauer et al. presented [10]. a new experiment-based quantum physics course for secondary schools, underlining the strong connection to contemporary applications of quantum optics in quantum information transfer The authors encourage the integration of single photon experiments and an interpretation of these experiments based on quantum theoretical fundamentals of light into the core concept of the course. As a consequence, one has to accept a significant distancing from Copenhagen-like quantum mechanics, such as atomic physics and stability of matter based on the discussion of Schrödinger's equation as an introductory concept.

As argued in [3], it is promising to minimize the number of axiomatic theoretical quantum features, absolutely necessary for achieving a deeper understanding of the quantum. Following this idea, we shall restrict ourselves to three basic traits: probability, superposition and interference (PSI). To allow for multivariant methods of access, we use the Dirac's bra-ket notation, complemented by a share of linear algebra formalism, which is adequate for school, and a geometrical pointer representation to get rid of the algebraic form of complex numbers directly using the imaginary unit, $i^2 = -1$

The central idea of a conceptual change from classical to specific quantum reasoning is introduced in Section 2. Here, we present the conversion of the central traits (Wesenszuege) of quantum physics [3] into an appropriate quantum reasoning and the underlying physical picture of the photon. Section 3 presents a study which provides empirical evidence of a substantial interest to physics teachers in true quantum physical experiments without classical bonds. The results of this study strongly reflect the consequences of this epistemic requirement on classroom physics and teaching methods. The perspective of the learners is reflected in Section 4, where the idea of fostering a specific readiness of students for quantum reasoning as a consequence of the engagement with a quantum physical key experiment (single photon states interacting with optical beam splitters) is introduced.

A comparative look at the current literature shows a surprising result: Though educational research on teaching quantum physics has enjoyed increasing popularity and revealed numerous (and various) concepts of how to teach quantum physics over the last 20 or so years, one finds a lack of empirical research into student learning barriers for teaching strategies intended for a conceptual approach to the quantum domain (for a comprehensive review, see [11]). The analysis given in [11] shows that the main learning barrier is the switch from quantum physics to physical reality. The basic idea of the analysis given in the present contribution is strongly related to the historical judgement that pre-QED (quantum electrodynamics) models of the photon are anachronistic, only seemingly providing a simplification at the expense of deeper understanding [12]. Engagement with the key experiment thus aims to get rid of misleading dualistic wave-particle models.

Due to pandemic measures, this study was shrunk greatly to an explorative interview study (a small sample of 36 students and a questionnaire shortened to three questions) to test the instrument which investigates the impact of the key experiment. Assuming that the background knowledge of the first-term students is not far from the typical knowledge of high school graduates, the explorative investigation was conducted with university physics students.

2. Quantum Reasoning: The Contemporary Model of the Photon

2.1. The Dusk of Dualism

From a contemporary perspective, one should state that interference phenomena of massive quanta such as electrons are the result of quantum theory based on the Schrödinger's equation and deBroglie wavelength, while the interference of light is a matter of classical Maxwell theory. Switching to the physics of the interaction between light and matter, the semiclassical theory, which introduces quantized matter and classical light fields, works remarkable well for a wide range of phenomena (photoelectric effect, Compton effect, spectroscopy, nonlinear optics, etc.). Subtle phenomena, however, force us to move away from the classical model of light. Known examples are the almost-zero-time delay between the emission of photoelectrons and the incident of the light in the photoelectric effect and the Casimir effect (this is due to vacuum fluctuations of the electromagnetic field) and the two-photon interference shown by the Hong-Ou-Mandel effect [13,14]. These phenomena cannot be explained by theories based on classical light.

Since the early days of quantum physics, the puzzling interference phenomena has led to dualistic hybrid models mixing features of particles and waves of physical items.

Quoting [13]: "Dual conceptions of light, as wave and particle, have co-existed since antiquity. Quantum mechanics officially sanctions this duality, and puts both concepts on an equal footing ... "

"Equal footing" seems to be a fairly cautious paraphrased way of getting rid of the dualistic "either–or" and "as-well-as" concepts of light. From this QED point of view, photons are neither waves nor particles. They are instead an intrinsic quantum optical entity that obeys the superposition principle. After interacting with an interferometer, it might be registered by a binary photodetector leading to a "click". As shown in the experiment (see Section 3), the accumulation of these detection events finally leads to an interference-like pattern of the counting rate (i.e., the detection probability). The idea of "equal footing" may be well suited to developing a mindset of the photon appropriate to get access to quantum reasoning beyond the classical horizon.

2.2. Conceptual Change to Quantum Reasoning

Today, the application perspective of the second quantum revolution fosters focussing on the conceptual understanding of quantum physics (e.g., [15]). These requirements are the background of the development of quantum physics as an often-compulsory part of upper secondary school curricula [3,16–21]. The move from traditional quantum mechanics to conceptual questions of quantum physics made students more attentive for fundamental differences between the explanations of the physical world and how physical reality is perceived [22]. Understanding phenomena of the quantum domain concerns the introduction of somewhat counterintuitive concepts such as probability, uncertainty, and superposition. A profound change in the mindset and corresponding educational concepts is needed [23–25]. More recently, some studies and educational proposals have given new stimuli to the discussion [26].

- Learning physics relies on conceptual understanding. It follows that misleading preconceptions or unwillingness to change a stable mindset inhibit physics learning.
- Learning quantum physics may be described as a change of reasoning not only scratching the surface but conceptually changing from classical reasoning to quantum rea-

soning, comparable to the changeover from the semi-classical Bohr's model to a full quantum model [27].

- Are there further theoretical approaches to understanding, and why do commonplace difficulties in quantum physics learning arise?

Some of these questions are touched on by diSessa's theory of Knowledge in Pieces (KiP) [28]. This theory describes learning physics as transforming the naïve sense of mechanism into a physical one. The term "sense of mechanism" refers to a cognitive heurism providing explanations for observed phenomena and arguments for the predictions of phenomena or possible events. The backbone of each sense of mechanism is a set of principles and laws, which have to be identified in the observation. The so-called "phenomenological primitives" (or "P-prims") are the superficial version of the physical sense of mechanism referring to the principles of children's concept of physics. P-prims are a result of the straightforward interpretation of the perception of everyday-phenomena. P-prims are primitive as they are self-explanatory and need no further justification. Thus, P-prims are strongly bounded to specific contexts, which allows different explanations for similar physical phenomena to exist [28].

To transform the naïve sense of mechanism into a deeper physical one, the role of p-prims has to be changed to a necessary set of components of a cognitive heurism to identify relevant physical laws and principles. For such a changeover of roles, it is necessary to decontextualize explanations. Decontextualization means that the same physical principles and laws always explain similar phenomena [28,29]. Because everyday contact with quantum physics is usually extremely rare, the existence of quantum physical p-prims are rather unlikely. Perhaps for this reason, learning quantum physics presupposes the transformation of senses of mechanism, from real-world-principles (such as local reality and determinism) to quantum principles (such as lost locality), an everlasting internal probability, a superposition principle and a specific Born's rule to re-enter the real world.

This paper deals with a role changeover from classical p-primes to a quantum sense of mechanism. This changeover is thought to be accomplished by decontextualizing a quantum reasoning based on PSI to explain the prototypical and thoroughly selected phenomena such as the single photon counter patterns from single photon experiments or from an electron diffraction tube.

2.3. Components of Quantum Reasoning

Following McNeill and Krajcik [30], three components constitute scientific explanation:

- The claim is an assertion or conclusion which addresses the problem or a phenomenon, to be explained. Quantum physical phenomena lead to quantum physical claims.
- The evidence supports the claim. It can be a set of scientific data, an observation or reading material. Thus, the evidence is always part of the real world. To support it, the evidence has to match the claim, according to the amount and quality of the data.
- The reasoning links claim and evidence, showing, why the evidence can be seen as supporting the claim, connecting the phenomena with the scientific laws and principles. Quantum claims need quantum features of reasoning.

Evidence in physics is always backed up by real-world experiments. It thus becomes understandable that reasoning which links the real-world evidence and phenomena from the quantum domain has to include a special design. Far from touching the deep logical questions of reasoning in quantum theory (e.g., [31]), the specific quantum reasoning proposed here turns out to be a mindset consisting of objects and elementary rules bijectively assigned to central terms of quantum theory (Wesenszuege [32]).

In this sense quantum reasoning is a mindset with the following characteristic features:

- It is designed to be consistent with a rigorously selected list of basic exclusive quantum theoretical axioms [33].
- It is well-suited to equally footing the features of quantum mechanics (states of massive quanta) and quantum electrodynamics (states of photons).

- It links claims from the quantum domain and evidence from the real world.

  Specific features of this rational are:
- Probability thinking, where properties of physical systems are completely incorporated into the quantum states. Principally different from classical physical states quantum states are just vectors in an abstract vector space. Each vector allows for the calculation of a probability to detect (or even measure) characteristics of the system. In quantum physics, we have to think in terms of probabilities derived from quantum states. One might state that there are no physical variables in quantum physics but solely probabilities for the assignment of specific values of the variable.
- Superposition thinking focuses on the superposition principle that is a core concept of quantum theory: The linear combination of quantum states again is a quantum state. Mathematically a superposition of two quantum states, $|\psi_1\rangle$ and $|\psi_2\rangle$, is nothing but the sum of these state vectors, weighted and phase shifted by specific amplitude factors, $|\psi\rangle = c_1 \cdot |\psi_1\rangle + c_2 \cdot |\psi_2\rangle$. The new quantum state $|\psi\rangle$ of the system must be distinguished from a *classical mixture* of ensemble states, where probability solely occurs because our exact knowledge of the components is incomplete. In the quantum superposition *each of the component states is always present*. In the weighted sum, $|\psi\rangle = c_1 \cdot |\psi_1\rangle + c_2 \cdot |\psi_2\rangle$, of the substates the coefficients c_1 and c_2 are closely related to an internal probability without any connection to the ability of the particular scientist.
- Interference thinking focusses on the detection of the superposition of states. A polarized spin-up-state $|u_z\rangle$ of atoms in the Stern-Gerlach apparatus confirms the superposition of a spin-up-state and a spin-down-state along the x-axis: $|u_z\rangle = (|u_x\rangle + |d_x\rangle)/\sqrt{2}$ directly measurable in the experiment. Generally, it is difficult to detect a superposition of states directly. To demonstrate a state superposition, $|\psi\rangle = c_1 \cdot |\psi_1\rangle + c_2 \cdot |\psi_2\rangle$, we are often forced to resort to demonstrating *interference effects*. Any phase difference between $c_1 \cdot |\psi_1\rangle$ and $c_2 \cdot |\psi_2\rangle$, temporally stable during the sampling time of the detector, may lead to interference patterns. One might say that there is no quantum interference without the superposition of states and that quantum interference is a safe indicator of a superposition of quantum states.

  See Appendix A for details of the pointer representation of quantum states.

3. Experimental Teaching in Classroom Quantum Physics

3.1. Theoretical Background

Science funding for digitalization and quantum technology (*The Quantum Flagship* [15]) is one of the most ambitious long-term European Commission research initiatives. The increasing popularity of quantum physics thus seems unsurprising. Our understanding of physics evolves hand in hand with advances in our understanding of the interaction of light and matter, as has been modelled since the 1950s by the modern version of the quantum electrodynamics. It seems, however, that the physical science of the last 100 years still represents a small share of classroom physics in German high schools, along with those in many other countries [10,11]. It looks as though quantum physics curricula worldwide are locked onto the historical development of quantum physics with some specific anachronisms (particle-wave-dualism).

The question of how to overcome these shortcomings leads to the corresponding teacher training requirements as follows.

- Following the major role of experiments in physics education, didactic aspects of quantum experiments and of science communication should play a substantial role in physics education [34–36].
- Interpreting quantum theory is a minor part of teachers' in- or pre-service training. Conceptual knowledge is often restricted to semiclassical concepts which ignore the basic concepts of quantum field theory, leading to the notorious problems which arise from the wave-particle dualism and localization of quanta.

- There is less information about students' views on quantum physics when compared to those of classical physics [37].
- The lack of pre-service teacher training on the educational and experimental efforts of quantum physics [38] is significant.
- Experiments such as single-photon and two-photon interferometry and current applications in quantum physics usually go beyond experimental and theoretical high school expertise.

To face these challenges, systematic training courses for teachers are required, which go beyond the semiclassical confinements as discussed by [9,39,40]. It is known that the courses have to comply with the real needs of teachers. First and foremost, the course must be in accordance with the particular curriculum, which assign quantum phenomena with single photons to a specific quantum reasoning (Table 1).

Table 1. Typical correspondence between quantum reasoning and quantum phenomena in hight school textbooks.

Phenomenon	Quantum Reasoning
Interference of single photons	Superposition of states
Double slit interferometer	Born's rule to determine the probability of an event
Mach-Zehnder-Interferometer	Nonlocality of single photon states in the interferometer
Blurring the interference pattern	Influence of measurement on system behaviour
Absence of coincidences at the optical beam splitter	Complementarity of anticorrelated events

3.2. An Empirical Study

The global aim of the project presented in this Section is to develop and offer a training course which fits these requirements. Questions such as the following turned out to be well-suited to structuring the complex research area.

- How do teachers handle the problems which arise from the epistemological clash in quantum physics at school without experiments?
- How do teachers assess their pre-service education regarding the requirements of quantum physics at school?
- How can we design a teacher training program that offers university-level experimental and theoretical backgrounds suited to the real needs of teachers?

A suitable methodological tool for exploring such a complex field of attitudes and knowledge is the established Delphi-method [41–43]. The Delphi study presented here has been through three rounds between May 2014 and July 2015 (see Table 2 [43]). The first round started with $N = 84$ study participants in mid-2014, the second round with $N = 54$ at the end of 2014 and the third round with $N = 70$ in mid-2015. The difference in the number of participants is mainly explained by nonattendance during the second round and an increasing attendance in the third. The size of the sample is small; this shortcoming is compensated for with the high quality of the answers.

Table 2. Details of a Delphi study.

Details of the Design	Example	Bundled Result
The questionnaire of the first round mainly consists of open questions	Which topics are challenging for your students and how do you or what do you need to master these challenges?	A typical statement: "Due to absence of experiments in school, most of principles of quantum physics can only be believed and memorized."

Table 2. *Cont.*

Details of the Design	Example	Bundled Result
The goal of the second round is to get access to the position of each participant to disputable answers, main attitudes and motivational topics identified in the first round. The questions are more focussed in comparison to round 1, without losing the open character to permit critical discussions.	Most of principles of quantum physics can only be believed and memorized due to absence of experiments. Please position yourself to the statement above and describe especially the role of simulations and animations.	A typical statement: "Simulations enable the discussion of the principles of quantum-physics in my class, but often my students don't trust in simulations."
The question immediately arising asks for the generalizability of the positions identified in the first two rounds. In the third round, critical opinions and disputes from the previous two round have been identified. The group of experts have been asked to rate these on a 5-point Likert-scale	Do you agree (1), . . . , disagree (5)? My students do not believe in the results of computer simulations in quantum physics.	Most of the participants disagree with the position "My students don't trust in results from simulations." 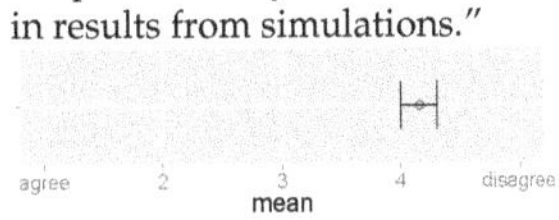

Condensing the results of the study into a course specification sheet, a few main implications for teacher training can be obtained (see Figure 1).

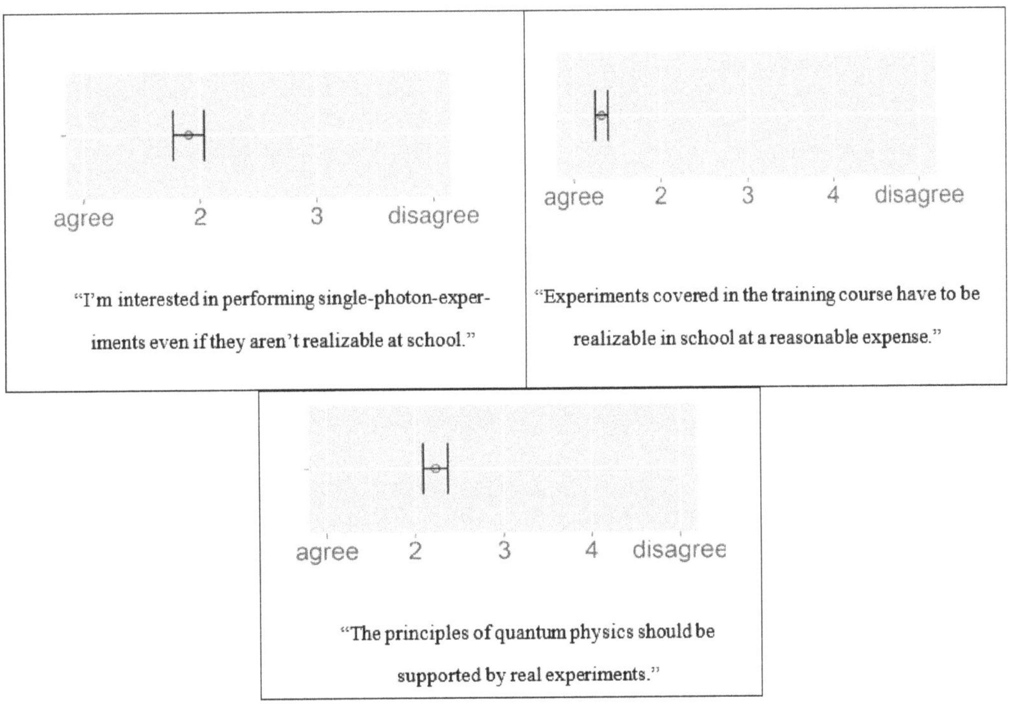

Figure 1. Teachers' basic needs [43]. The mean value and the standard deviation of the particular ratings is shown.

- Simulations and real (single-photon) experiments should be used in a mutually supportive manner.
- Focusing on experiments with interpretability does not depend on a formalism which goes beyond the scope of classroom mathematics.
- There should be an orientation towards principles of quantum reasoning.

The found results (Figure 1) well reflect the high level of classroom experiments:

3.3. A Quantum Physical Key Experiment

To address the consequences for teachers' training efforts, we conducted a study based on an experiment with heralded photons recently proposed for the preuniversity learning of quantum physics [34]. This experiment demonstrates two classically contradictory attributes of single photons: incompatibility of the beam splitter output states (absence of coincidences) and a $\cos^2$-dependence of the counting rate on the position of one of the mirrors in an interferometer set up (interference-like pattern of the counting rate). This emergence of maxima or minima of the measured photon counting rates (see Figure 2b) by changing the position of the mirror is quantum theoretically assigned to the difference of the phases of the substates of a state superposition: quantum interference.

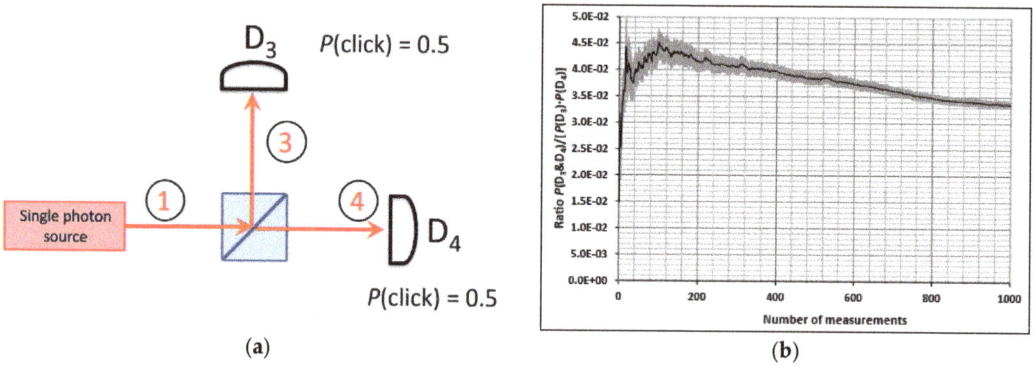

Figure 2. The number of coincident clicks of the detectors D_3 and D_4; (**a**) the setup (the blue square indicates the beam splitter, encircled numbers are assigned to the beam splitter modes; red lines indicate the direction of propagation of the light) and (**b**) experimental results: the mean value of 1000 measurements $\langle\alpha\rangle << 1$; the shaded area indicates the standard error. P denotes the probability of a detector click.

3.3.1. Part I: Specificities of Single Photon States

Consider the first part of the key experiment with single photon states (Fock states with $n = 1$) as the input of the beam splitter (Figure 2a). There are now two potential outcomes.

1. In any case, light behaves like classical electromagnetic fields and, therefore, the amplitude of the single-photon field (whatever that might be) will also just be split at the beam splitter. Depending on the physical properties of the light used in the experiment, clicks of D_3 and D_4 are more or less independent, the coincidences more or less accidental. For the joint probability of coincidences compared to the product of the marginals one finds $P(D_3 \,\&\, D_4) \geq P(D_3) \times P(D_4)$ [43].
2. The single-photon states of the light are made up of breakable particles. These particles behave like classical ones, and thus may burst, meaning that we have a finite probability $0 << P(D_3 \,\&\, D_4) \leq P(D_3) \times P(D_4)$ for coincident clicks.

Instead of the probability $P(D_3 \,\&\, D_4)$ itself, the ratio $\alpha = P(D_3 \,\&\, D_4)/[(P(D_3) \times P(D_4)]$ is shown. Classical fields (case (A)) would produce $\alpha \geq 1$. Breakable particles (Case (B)) would lead to $0 << \alpha < 1$. Figure 2b shows what happens instead. As can be seen, α and thus the mean value of the probability of coincidences is much smaller than expected from classical theories. The probability of coincident clicks from detector D_3 and D_4 almost vanishes and therefore too the probability of a splitting of the single photon state by the beam splitter.

3.3.2. Part II: Quantum Superposition

The interferometer setup is shown in Figure 3a. Again, the single photon state is the incident state of a beam splitter. Now however two mirrors are added (as shown in Figure 2a). After an argumentation based on inseparable single-photon-states, classical probability theory ignoring substates and their phases would lead to a constant probability, while the equally shared characteristic of the beam splitter would yield equal probabilities of 0.5 for D_2 clicking and for the possibility of the light being reflected back into the source, not dependent on the mirror's position.

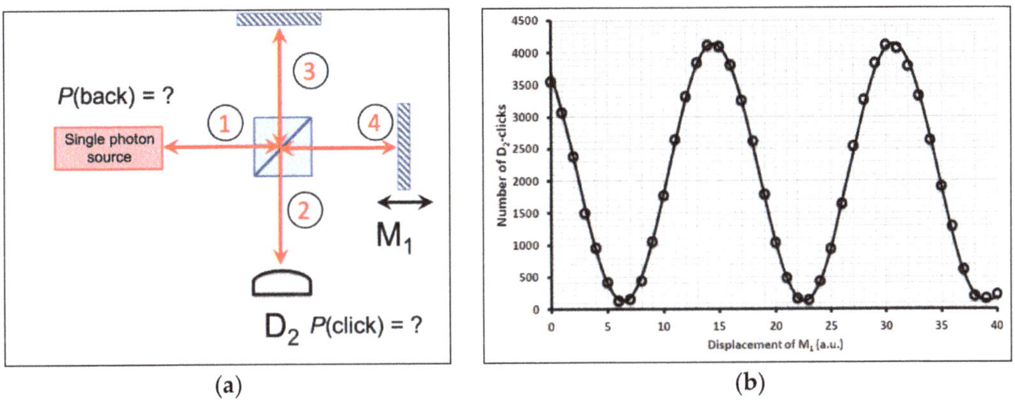

Figure 3. The number of single photon clicks of detector D_2; quantum interference: a displacement of the mirror M_1 produces a periodic variation of the counting rate even in the case of single photon states; visibility $V = 93\%$; (**a**) the setup and (**b**) experimental results.

The experimental result is completely different. Depending on the position of mirror M_1 the number of output counts changes. Periodically alternating, the light reaches D_2 or is reflected back into the source (the latter is not shown here). This pattern is interpretable as an interference (Figure 3b). There is no classical physical explanation because there is neither an electromagnetic wave to produce interference fringes nor more than one photon to allow for some inter-photonic interaction. Quantum theory, however, is well suited to solving this conflict. The quantum interference phenomenon shown here experimentally is a consequence of the superposition of two quantum sub-states, one of them the result of a transmission process, the other state corresponding to the reflection. The superposition state leads to interference fringes in the final probability.

Combining both experiments, a puzzling situation concerning the concurrence complementary phenomena occurs: The bare beam splitter itself shows a lack of the coincidences one would expect for unbreakable radiation elements. With mirrors added, the experiment shows an interference-like dependency on counting rates with the same beam splitter. A nonlocal interpretation (including the complete interferometer), based on quantum theory resolves the conflict, implying that this combination experiment possibly acts as a "door opener" to the quantum world. The quantum theoretical interpretation.

- does not concern the splitting of anything by beam splitters,
- does not concern paths of quanta,
- tells us everything about the probabilities of detection eventualities,
- makes use of the superposition principle together with Born's rule to explain quantum interference,
- allows us to find the quantum state prepared by a beam splitter as a nonlocal superposition of single photon substates, and
- encourages us to rename beam splitters "quantum state preparators".

The key experiment might help to satisfy physics teachers' needs, if they are looking for real experiments which demonstrate true quantum physics which well demonstrate the failure of classical theory.

4. Impact of a Key Experiment on Quantum Reasoning

4.1. Research Goal

As has just been pointed out, the key experiment can be explained by using the basic components of quantum physical reasoning PSI. On the other hand, conducting and discussing the two parts of the experiment separately and independently risks stabilizing the well-established and attractive semiclassical wave–particle dualism. Closely combining these experiments, the wave-particle dualism does not well explain the concurrence of photon anti-correlation and interference in a single experiment. The experiment thus might function as a door opener to the quantum domain, motivating the development of a quantum reasoning which relies on PSI [8,9].

Closely related to the present contribution recent studies have reported on approaches to quantum physics that rely on student insights into the general relevance of quantum technology for everyday life [44]; that rely on the analysis of typical patterns of student difficulties in modern quantum physics in schools [45]; and that rely on a specific model of a learner's way of understanding quantum physics [46]. While real quantum experiments with single quanta have been available for many years, they are still not yet practicable in schools [34,47]. It therefore comes as no surprise that (at least to our knowledge) the impact of quantum key experiments on learning about quantum physics remains unknown. To obtain greater insight into the experiment's impact on students' readiness to use quantum terms in their interpretation of the experiment, it would be of interest to investigate.

3. in which regard the students' lines of argumentation move from classical reasoning (either-or) or a semiclassical dualism (as-well-as) to quantum reasoning (PSI);
4. which types of specific (quantum) reasoning (PSI; neither-nor) can be found for the explanation of quantum phenomena such as the key experiment after students had engaged with the key experiment.

4.2. Method and Sample

Unfortunately, due to the pandemic measures, it was not possible to get in touch with high school students. As a preliminary step, in order to test the pool of items and to see whether the idea functions in principle, a reserve sample was used.

For this purpose, 80 physics students (first term at university) in a one-group, pre-design and post-design were engaged with the key experiment. To gain empirical access to the experiment's influence on student reasoning, the empirical construct "use of quantum reasoning" was improved as explained above (usage of the PSI-line of argumentation) and a mixed-format test and an additional semi-structured interviews with a subgroup of 36 students were used. These interviews focussed on student reasoning for the explanation of classical contradictory phenomena of the key experiment. The findings are the topic of this section.

The interviews with a mean duration of 14.3 min (pre) and 12.0 min (post); standard deviation, SD = 3 min in both cases, were digitally conducted and recorded via an anonymous video conference. An interview guide with three tasks was developed in order to provide a basic structure; the guidelines were prepared according to [48] (see Table 3):

- Task 1: Basic understanding and description of the term photon.
- Task 2: The explanation of the absence of coincidences in the experiment with the naked beam-splitter.
- Task 3: The explanation of the counter pattern in the interferometer set up.

To visualise the key experiment, appropriate figures were provided in Tasks 2 and 3, displaying the absence of coincidences at the beam-splitter and the counter pattern with mirrors added. For students who were unable to answer the questions potential answers are made available, taking into account known misconceptions and appropriate responses.

As taken up in in the conclusion, the interview setup, separating the two components of the experiment into two tasks, was at risk of stabilizing the concept of the classical particle.

Table 3. Guidelines for the interviews.

Question	Hints and Instructions	Additional Material
Task 1 In this interview I would like to talk with you about the photon and its behaviour. Please describe your understanding of the photon. Explain: What is youconception of photons?	What is your conception of particles/waves?	Answer options: A photon is an undividable energy quantum of the light field: 1. a light particle, which is moved by the enveloping light wave. Depending on the experimental setup, the behaviour of the photon or the light wave occurs (the classical either–or argument). 2. existing as wave and particle concurrently. Depending on the experimental setup, it occurs as a wave or as a particle (the dualistic argument). 3. neither a particle nor a wave, but it behaves sometimes similar to a wave or a particle (the quantum neither–nor argument).
Task 2 Single photons interacting with a beam splitter: Report the results of the experiment. Explain the results	Hints regarding relative frequency of events: What's about features of particles? How does the result fits in with your concept of photons? What can be said about the state of the light at the output of the beam splitter? Try to give an explanation in terms of two different output modes of the beam splitter	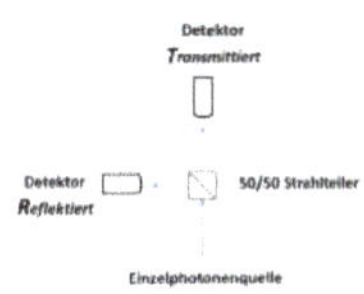 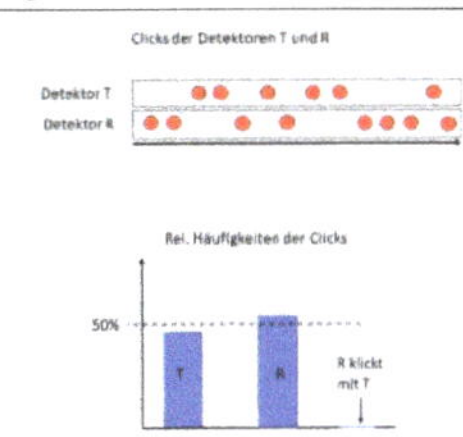
Task 3 Single photons interacting with the Michelson interferometer: Report the results of the experiment Explain the results	Questioning the impact of the mirrors: What's about features of waves? How does the result fits in with your concept of photons? What can be said about the state of the light at the screen? Try to give an explanation in terms of the superposition two different output modes of the beam splitter.	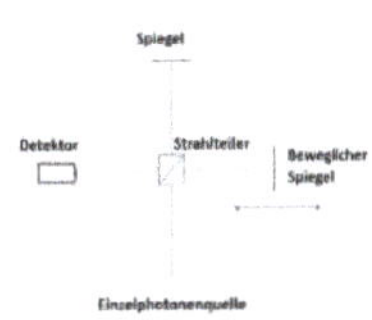 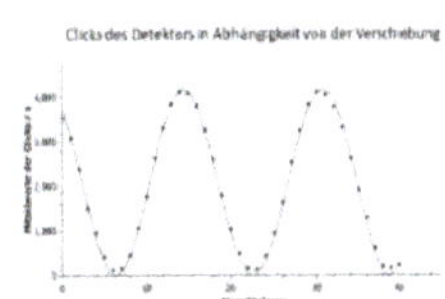

Following Ref. [49], students' conceptions can essentially be placed into three categories: classical, dualistic and quasi-quantum physical and assigned to each task. Table 3 lists the questions of the interview guide translated from German and drastically shortened. For a detailed version, see Ref. [26]. The interview study was conducted in German. Semantic deviations from the original questionary due to the translation may have been unavoidable.

To evaluate the coherence of the student responses, the students were advised to link the results from Task 2 and Task 3 to the results from prior tasks.

4.3. Analysis

The interviews recorded were transcribed and subsequently paraphrased, thus reducing the number of utterances to the relevant aspects (i.e., a student's reasoning for the explanation of quantum phenomena). To analyse the students' responses, the method of structuring qualitative content analysis [50] was used, because it allows us to identify and conceptualise contend-related aspects in the material. Based on this conceptualization, the method allows for a systematic description of the material with respect to these content-related aspects [51]. The use of a deductively-inductively derived coding system is the method of choice for this analysis. To create this system a basic set of categories was obtained from the literature and enriched with sub-categories derived from an analysis of the interviews [50]. As basic categories, classical, dualistic and quasi-quantum categories were chosen.

In this way it was possible to assign the students' responses to categories even if they were given in more complex response patterns (see Table 4). To illustrate this process let us give an example for a typical dualistic reasoning:

Table 4. Categorial system used to analyse the interview data; the first column refers to the three basic tasks, the second and third columns show main categories and subcategories derived from paraphrasing the individual responses, paraphrases are described in the fourth column, the last column gives an example.

	Main Category	Subcategory	Description	Example
Describe the term photon	Classical		The photon is described as classical particle or is associated with the properties of a classical particle	Photons are light particles, energy packages of electromagnetic radiation. Photons are portions of light (h × f), which can be modelled in different ways, but mainly it is modelled as classical particle.
	Dualistic		Dualistic description of photons: Hybrids of waves and particles, like for example: a wave-particle, a particle with wave properties, complementary behaviour of photons	Light is composed of photons, which are waves as well as particles. Depending on the experiment one gets the one or the other. A photon is something small, which is a wave (wavy path) as well as a particle (straight direction of propagation). The experimental setup decides which property of a photon can be observed.
	Quasi-Quantum		An object which has wave like and particle like properties, but is neither a wave nor a particle; Attribution of probabilistic behaviour/ probability amplitudes	A photon is an object which has properties of a wave and a particle, but is neither a wave nor a particle. A photon is a small energy package, which can't be described by classical physics. It can be described by using probability amplitudes
	Other		Cannot be categorized	Photons are positive charged particles, which work as a current of light
Optical beam splitter	Particle		Explanation of the experimental results by using particle-based reasoning	
		Classical	Using properties of a classical particle (e.g., realistic arguments)	In this experiment a single photon can't be divided into two halves, because it can be at only one single position. At an optical beam-splitter, photons were reflected or transmitted with a probability of 50%, depending of the amplitude of the transporting wave. Due to particle properties, the photon can't be divided. With a probability of 50% photons were reflected or transmitted, but they will never be divided, because the photon has to choose one path. It cannot be said anything about wave or particle properties.
		Dualistic	Due to the experimental setup, the photons occur as particles.	Because the experimental setup allows to measure the photon's location it demonstrates particle behaviour.

Table 4. *Cont.*

	Main Category	Subcategory	Description	Example
		Quasi-Quantum	Photons behave like particles, but they are no particles	Due to the particle characteristics, the photon can't be divided into halves like a wave. Thus, we get "either-or"-results.
	Energy Quantum		Explaining the results by the indivisibility of the photon's energy.	With a probability of 50% single photons will be transmitted or reflected at the optical beam-splitter, but neither divided in two halves, because they are indivisible energy quanta. The photon decides whether to be reflected or transmitted in probabilistic ways.
	Probability		The experimental results will be explained by using a probabilistic reasoning.	
		Non-localisation	The experimental results need a probabilistic explanation, because the photon position is not determined until the photon's detection.	The path taken by the photon, is unknown until the photon's detection. However, the detection is arbitrary, with a probability of 50%. Single photons will be reflected or transmitted with a probability of 50%, but neither divided into two halves. This looks like a particle property, but the behaviour can only be described by probability amplitudes. Until the photon's detections it is not determined whether the photon is reflected or transmitted.
		(Non-localisation + superposition	Explaining the experimental results by describing the final state as a superposition of the substates reflected and transmitted.	Photons will be reflected or transmitted with a probability of 50%. Until the photon's detection the photon's path is not determined, but it can be described as a probability amplitude. By detecting the photon, the superposition of probability amplitudes will be destroyed.
	Choice		A provided explanation is chosen by the student.	Choose Explanation1, because the photons show up as indivisibleenergy quanta.
	Other		The explanation cannot be sorted into one of the categories.	No explanation, but the photons are distributed randomly.
Interferometer	Wave interference		Explaining the results like the interference of waves.	
		Wave characteristics	Due to the wave characteristics of photons, the interference occurs.	Interference can be observed in this experiment, because of the wave characteristics of photons. By moving the mirror, a difference in the path length is realized. Particle characteristics would lead to a constant number of counts.

Table 4. *Cont.*

Main Category	Subcategory	Description	Example
	Dualism	The photon shows up as a wave/the experimental setup determines the photon as a wave	Intensity minima and maxima can be observed because of the interference. Due to the superposition of wave and particle, unbreakable particles show up as waves in this experiment, because the experiment demonstrates the interference as a classical wave property. The diagram shows a sinusoidal click distribution, because the wave properties are observed in this experiment. Thus, the photon cannot be regarded as a particle/localized object. It is in a subordinated state. Depending on the superposition of the amplitudes, constructive or destructive interference can be observed. It becomes understandable, how the photon's properties are determined by the experimental setup.
Probability interference		The observed interference is a probability interference.	
	Non-localisation	A probability interference is observed, because the photon's position cannot be determined. → Basic probabilistic reasoning	In this experiment constructive and destructive interference can be observed, because reflected or transmitted photon 50% probability is reflected on the beam-splitter again, by the mirrors. Thus, it is impossible to determine whether the photon is reflected or transmitted and probabilities will interfere. This experiment demonstrates a wave characteristic.
	Superposition of probability amplitudes	The photon must be described by probability amplitudes and the superposition of the probability amplitudes causes the interference. → Advanced probabilistic reasoning	In this experiment the probability amplitudes of the both, possible paths the photon could take, are superimposed. By moving the mirror, a difference in the paths is realized and the inference pattern changes from constructive to destructive interference of the probability. Thus, a wave characteristic is attributed to the photon. In this experiment constructive and destructive interference can be observed, because the probability distributions of the both possibilities interfere. For each possibility, reflection or transmission, a probability distribution exists, which can be regarded as waves in the arms of the interferometer. Thus, by moving the mirror a phase difference is created.
Wave behaviour		The experiment demonstrates the photon's wave properties but no interference	More or less photons are detected, because depending on the distance between mirror and beam-splitter more or less photons can be registered. This must be the wave property, because a particle will be detected with a property of 50%.
Other		The explanation cannot be categorized.	Not an explanation, but the chose option 2 or 3. The experiment allows no opportunity to talk about the indivisibility, because it can't be measured whether the photon is reflected or transmitted. And the energy of the photon changes by increasing the distance between beam-splitter and mirror, due to air friction.

Table 4. *Cont.*

Main Category	Subcategory	Description	Example
No explanation		Student choose an option without any further explanation.	

This interview can be paraphrased: Photons are light particles, transporting and releasing energy (energy quanta). However, they are also electromagnetic waves and have characteristics of waves.

This type of content-related analysis always risks subjective rating, leading to reliability issues of the results. To adhere to due diligence obligations, we checked the categorization by two independent raters. Due two different priorities (single statements vs. more complex patterns) we found Cohen's kappa [50] M(κ) = 0.6/SD(κ) = 0.2, mean valued over all three tasks/pre and post. After clarifying the reason for the discrepancy, we reached full interrater agreement. A piloting test of the questionnaire ensured the students ability to understand the technical language and to edit the questionnaire.

Student: Eh, I see the photon as a tiny particle which transports a portion of energy as a light quantum, thus, a particle, energy and it is also an electromagnetic wave which has particle properties.

Interviewer: Ok, now you have mentioned three different aspects. I want to take a closer look at them. You said, photons are something like particles, did not you?

S: Yes

I: Please explain what it means to you.

S: As I already explained, (they are) portions of energy, light as a particle. Light may be presented as a particle, able to release energy just like a particle. e.g., as in the case of the photoelectric effect, where the light particles transfer their energy to the metallic plate and the electron will be emitted.

I: Ok, as a second aspect you mentioned light quanta and as a third aspect you mentioned electromagnetic waves. How would you further specify these aspects?

S: Ehm, light as an electromagnetic wave has wave properties, like for example diffraction on surfaces or slits or light refraction.

I: If I understood you correctly, light has properties of waves as well as properties of particles?

S: Yes, that is what think about it.

I: Please try to go into greater detail.

S: Do you mean the wave-particle dualism of light, that it can be both, particle and wave.

I: It is irrelevant what I am thinking about it. I just try to find out what you think about it (laughing).

S (laughing): I think there is a duality and light have both features, features of particles and waves as well.

I: Simultaneously?

S: Yes, simultaneously.

4.4. Results

4.4.1. Description of a Photon

During the "pre"-status, more than half of the students describe a photon as a dualistic wave-particle-hybrid (Figure 4): Photons are light particles, which transport and emit energy. However, they are also concurrently electromagnetic waves.

37% of the students described a photon in a quasi-quantum physical way: Photons are small bundles of energy, which cannot be well categorised as waves or particles. Depending on the situation they show wave or particle behaviour.

9% of the students described a photon as a classical particle. A wavy behaviour emerged as a feature of larger samples: Light is composed of photons, which are portions of energy. The photons are considered as particles. However, a bunch of photons will behave like a wave. Depending on the experimental situation I'm observing the characteristics of the wave or of the particles.

For the post-interviews an increase in the quasi-quantum physical description (53%) and a decrease in the dualistic description (33%) could be recognized. The classical description of a photon has been stabilized; we found an increase for the classic position (+2 people/5%).

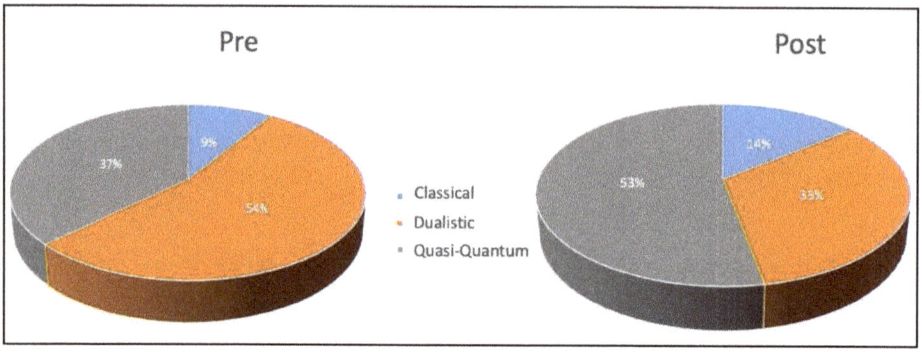

Figure 4. Students' understanding of a photon, before ("pre") and after ("post") they were engaged with the experiment.

4.4.2. Beam-Splitter Experiment

For the beam-splitter experiment (Figure 5), one can see that in the pre-interviews, 50% of the students preferred a particle argument, in which students talk about classical particle, dualistic particle occurrence or something like a particle behaviour of a rather obscure object; 25% came up with a probability argument, such as: Photons are indivisible energy quanta, for which only probability distributions can be formulated behind the beam-splitter. Only the measurement will lead to a certain result.

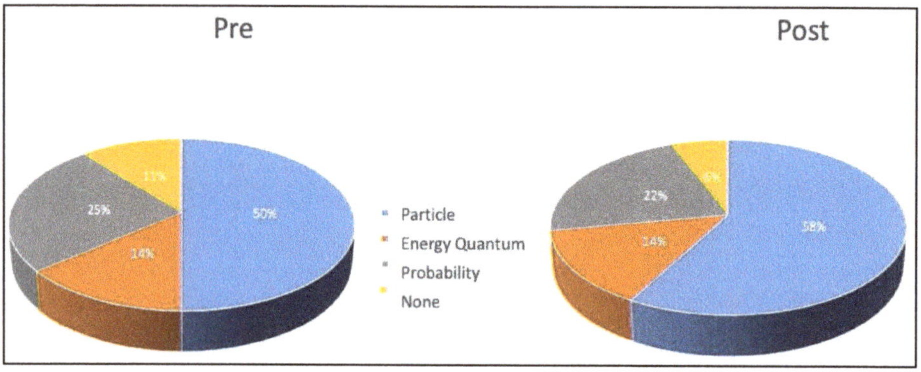

Figure 5. Students' reasoning for the explanation of the beam-splitter experiment, before ("pre") and after ("post") they were engaged with the experiment.

One student used the principle of superposition (subcategory of probability arguments).

A group of 14% used energy quantization as an argument to explain the results, such as: Photons will be reflected or transmitted with a probability of 50% and will produce a click in one of the detectors, with a certain probability. The photons do not hit both of the detectors, because they are indivisible energy quanta.

A group of 11% cannot be assigned to any category, because either the argumentation was entirely wrong (like one person who mixed-up photons and protons).

For the post interviews, a slight decrease in the probability arguments (−1 person using superposition) and a significant reduction in the non-categorizable paraphrase can be seen, while there is no change in arguments with quantized energy. Here, we find a visible increase in the particle position to 58% (+3 persons).

To get deeper insight into the different particle arguments, Figure 6 shows the distribution in classical, dualistic and quasi-quantum physical particle arguments in absolute numbers. In the pre-interviews four people used a classical particle argument: Photons will be reflected or transmitted with the same probability (50%).

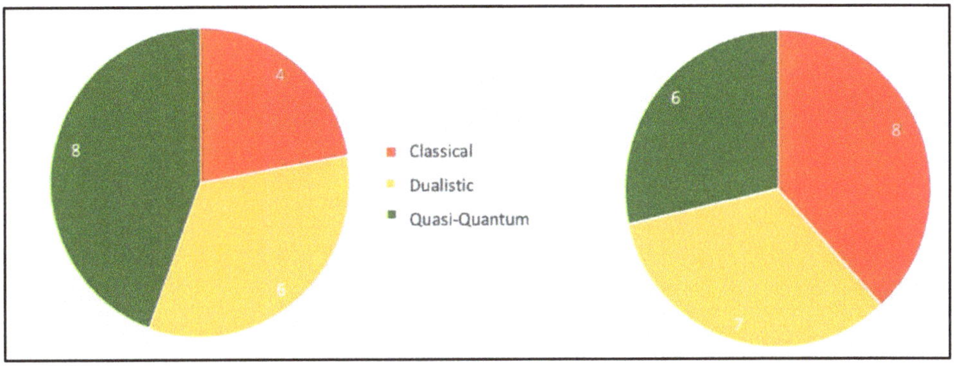

Figure 6. Particle arguments in detail (**left**) before the students were engaged with the experiment and (**right**): after the students were engaged with the experiment. The absolute numbers of the students are indicated.

This fosters the idea of photons as localized hard particles: Because particles could only be at one point at the same time, waves, however, could be transmitted and reflected concurrently.

Additionally, eight people used a quasi-quantum particle argument: The experiments demonstrate the elementary particle character of a photon. This is fostered by the inseparability.

Six persons used a dualistic argument: The photons will either be completely reflected or transmitted, but not divided into halves, because then photons occur as particles, which can only be completely reflected or transmitted.

For the post-interviews, an increase in classical particle arguments (+4 persons) and the number of dualistic-particle arguments (+1 person) can be seen, while the number of quasi-quantum-particle arguments decreased (−2 persons).

4.4.3. Michelson Interferometer

For the Michelson interferometer, one can see that 42% of the students who use wave arguments for their explanation (which can be specified as 53% of the group), say photons behave like waves due to a basic wavy character (Figure 7):

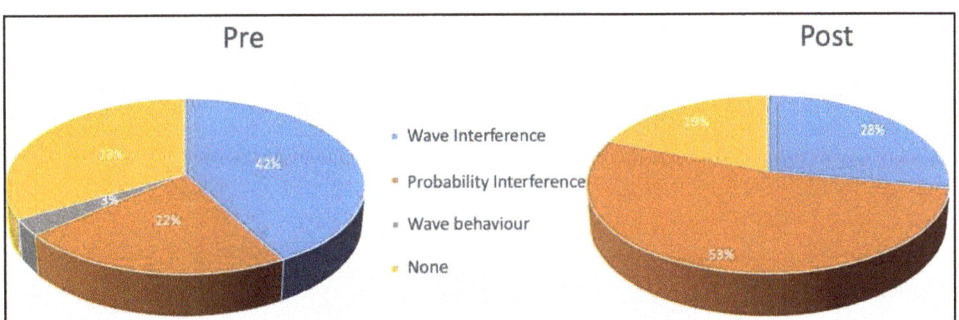

Figure 7. Students' reasoning for the explanation of single photon interference.

Destructive and constructive interference can be observed in this experiment, due to the wave character of photons. Moving the mirror produces a phase difference. A particle character, however, would produce a constant rate of counts.

A total of 47% use a dualistic reasoning: Constructive and destructive interference can be measured, because the experimental setup does not allow us to measure whether the photon is reflected or transmitted. The wave character therefore occurs. By moving the mirror, the waves were displaced against each other and as a result the detector measures the photon's wave.

A quarter of the students used probability arguments, while half of the group has already used the principle of superposition: In this experiment the wave behaviour of photons is demonstrated. Against classical waves, however, photons are not divided into two halves at the optical beam-splitter, but there is a superposition of all the possible paths a photon could take to the detector. Moving the mirror now produces a phase difference in the superposition and a single photon cannot be detected. Thus, more photons are need.

Nevertheless, 13 persons/33% of the students do not recognize an interference pattern, though at least 3% argued, that the result must be a wave behaviour: The experiment demonstrates the wave character of photons, because depending on the mirror's position, more or fewer photons will be detected. Due to the photon's position on the light wave, the reflexivity of the beam-splitter will change.

When analysing the post interviews, it can be seen that the amount of non-categorized paraphrases (19%) and wave arguments decreases (28%), while probability arguments increase (53%). Going into further details, it can be seen that 76% of these students used the superposition of probability amplitudes for their explanation. Nevertheless, 19% of the students interviewed did not recognize the interference pattern, although they analysed it in the key experiment.

4.5. Conclusions

The analysis of the interviews shows that engaging students with the key experiment seems to challenge dualistic conceptions/reasoning. For the photon description, an increase in quasi-quantum physical descriptions and a decrease in dualistic descriptions can be seen. Nearly the same can be recognized for the explanation of the interference.

By contrast an increase in a classical particle conception and reasoning can also be observed. The beam-splitter part of the key experiment seems to foster particular classical reasoning. Here, an increase in classical particle arguments can be seen (+4 persons), while the use of the principle of superposition is slightly increasing (+3 persons). It can therefore be concluded that the students did not understand the idea of superposition. This conclusion is fostered by the explanation of the interference. Here, 39% of the students explained it by using the superposition of probability amplitudes (~6% in pre-design). It can thus be concluded that the students deem superposition necessary only in the context of interference, but not as a fundamental principle for the explanation. A deficiency in the organisation of the interview study may additionally support this result. Due to the fact that the two parts of the experiment could not be presented as a closed unit, the particle concept has been supported.

4.6. Limitations

The analysis of the interviews gave valuable insights into the key experiment's impact on the change in rationales for the interpretations of a quantum phenomenon, uncovering the possible need for change of the concept and the organisation of the study. Some limitations of the study should be underlined.

- The size and composition of the sample of the present test are not satisfactory for obtaining robust results which answer the research questions. However, we received strong suggestions regarding a redesign of the questionnaire (items concerning the argumentation with the superposition of states must be improved).

- Perhaps the most important limitation is the design of this study. Because of the missing control group, the observed effect cannot be attributed to the treatment without uncertainty. Thus, only evidence-based suppositions about the key experiment's impact on students learning quantum reasoning can be derived [52]. On the other hand, one group designed studies seem to be suitable for gathering fruitful hypotheses about a treatment's effect and can therefore be starting points for subsequent studies (see, e.g., [53]).
- Due to pandemic conditions requiring social distancing, the students had no chance to really engage with the experiment. Instead, they were reliant on a digital version of the set up. It is to be expected that the impact of the experiment was drastically lowered due to this shortcoming. For this reason, its comparability with other studies is assumed to be quite limited and thus omitted in the present paper.
- Finally, the sustainability of the effect can be assumed to be low, due to the singularity and shortness of the intervention (length ~4 h). However, the aim was to get insights into the experiment's effect on learning quantum reasoning. Based on these results, implications can be made for teaching strategies based on key experiments, especially for gaining a more sustainable effect. More research in this field is necessary.

5. Discussion

Quoting [54]: "The universe revealed by modern research on the foundations of quantum mechanics is a strange and wonderful place ... As a matter of fact, our suspicion is that this [how to explaining it, Author 4] will prove to be impossible. For surely, to explain something is to reduce it to what is already known. But it may turn out that we will never be able to reduce the quantum universe to our customary ways of thinking. Perhaps we will have to adjust our ways of thinking to it. Perhaps years from now, people will think in new and unfamiliar ways, ways in which the quantum universe is no longer a challenge, but rather simple everyday reality."

In this paper, we present a rigorous formulated concept for an access to fundamentals of basic quantum phenomena. The argumentation solely relies on a quantum reasoning directly corresponding to a contemporary quantum traits approach to quantum physics [10]: probability, superposition and interference. The mathematics could be restricted to secondary school linear algebra and a straightforward pointer representation for the addition and multiplication of complex numbers. Our concept makes no use of classical argumentation and thus might show that there ought to be no resemblance between the classical and the quantum domain. We were able to show that, if one wants to understand quantum physics one has to be ready for a lane change [8] and that it is not impossible to make that change.

Author Contributions: Conceptualization, R.S. and S.W.; methodology, K.-A.W., M.W. and S.W.; validation, K.-A.W., M.W. and S.W.; formal analysis, R.S. and S.W.; investigation, R.S., S.W., K.-A.W. and M.W.; resources, R.S..; data curation, K.-A.W. and M.W.; writing—original draft preparation, R.S.; writing—review and editing, R.S.; visualization, R.S.; supervision, R.S.; project administration, R.S. and S.W.; funding acquisition, R.S. All authors have read and agreed to the published version of the manuscript.

Funding: We acknowledge financial support from Deutsche Forschungsgemeinschaft/DFG through collaborative research centre/CRC 1227 (Designed Quantum States of Matter/DQ-*mat*), project Ö, Leibniz Universität Hannover, Hannover, Germany.

Institutional Review Board Statement: Ethical review and approval were waived for this study, due to the fact that the study was in accordance with the Local Legislation and Institutional Requirements: Research Funding Principles (https://www.dfg.de/en/research_funding/principles_dfg_funding/research_data/index.html, (accessed on 5 September 2022)), and General Data Protection Regulation (https://www.datenschutz-grundverordnung.eu/wp-content/uploads/2016/04/CONSIL_ST_5419_2016_INIT_EN_TXT.pdf (accessed on 5 September 2022)).

Informed Consent Statement: Informed consent was obtained from all subjects involved in the study.

Data Availability Statement: The data can be obtained upon request from one of authors (M.W.).

Conflicts of Interest: The authors declare no conflict of interest.

Appendix A. Pointer Algebra of Quantum States

First let us emphasize that our analysis accepts one limitation that is apparent in high school physics textbooks: The difficile epistemic differences between quantum mechanics (physics of atoms, molecules, solid state bodies) and quantum field theory (physics of many body systems, physics of creation and annihilation of quanta) are mainly ignored. Instead, we have restricted ourselves to an equal footing strategy, relying on a set of basic theoretical traits similarly valid in any case. As explained in Section 2, we have chosen the triple probability-superposition-quantum interference. For more details of quantum theory and a representation of the basic ideas and for the mathematical formalism we refer to standard quantum physics textbooks [55–57].

Appendix A.1. The Geometrical Interpretation of the Phasor $q(\phi)$

The phasors which occur as q-factor constitute a concept of invaluable importance in science and technology. Figures A1–A3 picture a geometrical representation of electrical oscillations by a rotating phasor. Figure A2 demonstrates how to get the usual trigonometrical term from the projection on the abscissa. The electrical field strength of an oscillating field, $E_0 \times \cos \omega t$, can be derived from two opposite rotating phasors, E^+ and E^-:

$$E(t) = E_0 \cdot \cos \omega t = \frac{1}{2}(E^+ + E^-). \quad \text{(A1)}$$

To facilitate the solution of problems in the physics of oscillations and waves, phasors are used instead of the trigonometrical functions. A generalization of the idea draws phasors of length $|q| = 1$ and a phase angle ϕ (Figure A3), depending on the system under observation. It is straightforward to deduce calculation rules for $q(\phi)$ with $|q(\phi)| = 1$ (e.g., from the definition $q(\phi) = \exp(i\,\phi)$) (Table A1).

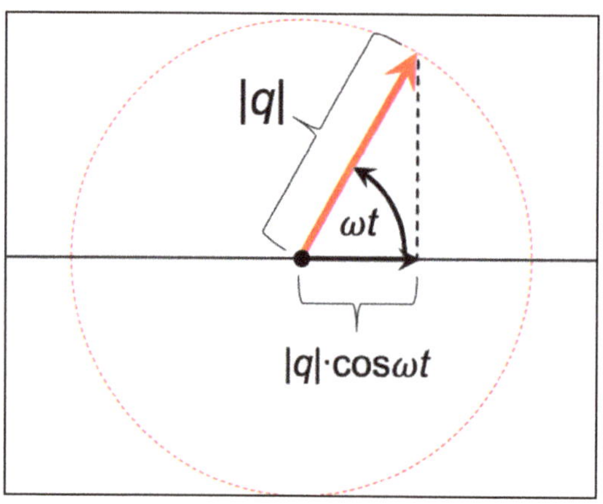

Figure A1. Phasor representation of an oscillation $q{\cdot}\cos\omega t$.

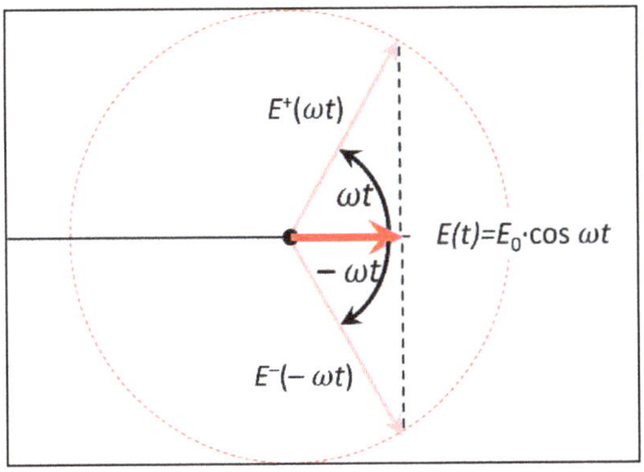

Figure A2. Phasor representation of an oscillating field.

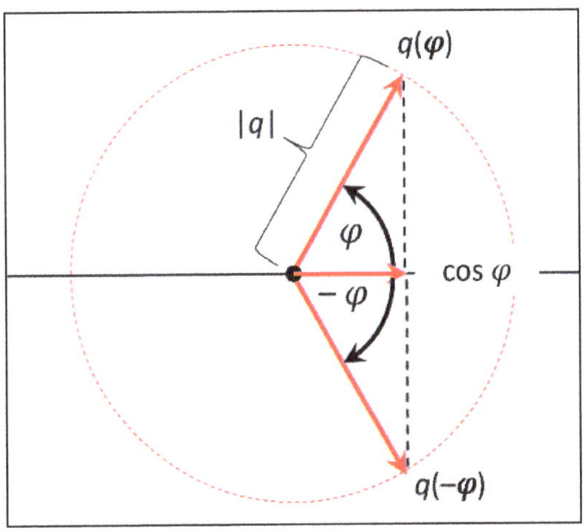

Figure A3. Generalized phasors.

Table A1. Calculation rules for the phasors $q(\phi)$.

Operation	$q(\phi)$-algebra
Multiplication of $q(\phi)$ (geometrical interpreted as a pointer rotation)	$q(\varphi_1) \cdot q(\varphi_2) = q(\varphi_1 + \varphi_2)$; $q^2(\varphi) = q(2\varphi)$
Some special values of $q(\phi)$	$q(0) = 1$, $q(\pi) = -1$
The absolute squared value	$\lvert q(\varphi)\rvert^2 = q(\varphi) \cdot q(-\varphi) = q(0) = 1$
The values for negative phase angles	$q(-\varphi) = 1/q(\varphi)$
Addition of $q(\phi)$	$q(\varphi_1) + q(\varphi_2) = 2 \cdot \cos((\varphi_1 + \varphi_2)/2) \cdot \cos((\varphi_1 - \varphi_2)/2) \Rightarrow$ $q(\varphi) + q(-\varphi) = 2\cos(\varphi)$ and $q(\varphi) - q(-\varphi) = 2 \cdot q(\pi/2) \cdot \sin(\varphi)$

Appendix A.2. Pointers and Quantum States

In the quantum domain the factor $q(\phi)$ carries the complete phase information of the quantum state. As stated above we have $q(\phi) = \exp(\mathrm{i}\ \phi)$ and thus $q(\phi)$ can be viewed as the algebraic version of the pointer representation of $\exp(\mathrm{i}\ \phi) = \cos\phi + \mathrm{i}\cdot\sin\phi$. In this sense Equation (A1) is close to Feynman's pointer representation of quantum electrodynamics and has proven extremely useful for an educational approach to the physics of the interaction of light and matter [58]. Furthermore, the angle ϕ makes it possible to give an illustrative interpretation of the quantum theoretical scattering amplitude of photons by atoms. The probability amplitude of detecting the photon by a detector placed at distance d from the atom and with c as the velocity of light ([55], p. 158):

$$f(d,t) \propto \exp\left[-\mathrm{i}\omega\left(t - \frac{d}{c}\right)\right]. \tag{A2}$$

Here, we demonstrate the application of pointers for our key experiment.

- Quantum states are represented by pointers. Positive real numbers by pointers with phase 0 (3 o'clock position) negative ones with phase π. Phases between 0 and π belong to numbers with an imaginary part.
- Here, photons are basic energy quanta populating the energy states of physical systems. The evolution of the quantum states of photons is modelled by the rotation of the pointers (Equation (A2)).
- The length of the pointer is a measure of the expected number of clicks of the detection set up (considered proportional to the number of incident photons); the phase of the state is identical to the phase angle of the pointer.
- Pointer rules (how to add and multiply them) and the calculation of the area of the square, sided by the pointer length, transform Born's rule (see [59]) into the pointer domain.
- The pointer length corresponds to $\sqrt{p}$. The square area sided by the pointer length thus gives p, the probability for a de-tector click.
- The algebraic representation of the pointer is given by $\sqrt{p} \cdot q(\varphi) \rightarrow \text{pointer}(|\psi\rangle)$

Appendix A.3. Counting Single Photons—Coincident Clicks

One of the most important optical components of quantum optics is the optical beam splitter (Figure A4). Textbooks on quantum optics discuss it widely (see, e.g., [60]). Here, we underline the "key model perspective": Optical beam splitters can be understood with our reduced model and suitable for performing experiments to uncover the quantum character of single photon states.

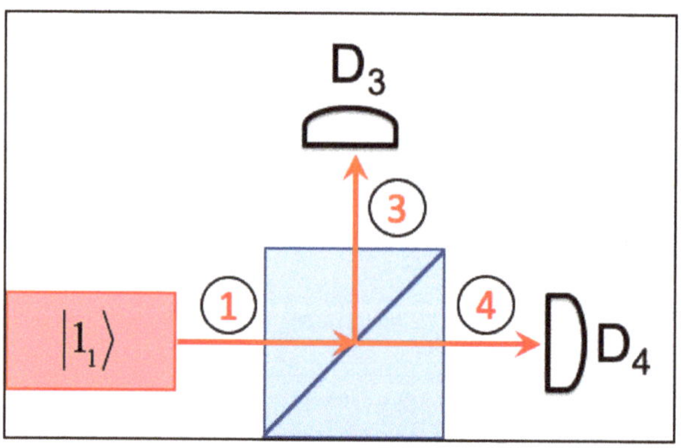

Figure A4. The optical beam splitter. $|1_1\rangle$ describes the Fock state at the entrance mode of the beam splitter. See Figure 2a for more details.

The binary detection scheme is realized, using quite low intensities of light. The binary detectors will only produce a voltage pulse, a "click" or not. For the analysis one usually estimates click probabilities from the numbers of clicks detected, N_i. The temporal length of a total measurement cycle may be denoted by T (e.g., 1 s); Δw (e.g., 5 ns) giving the temporal width resolving the minimum time between two different clicks (the so-called coincidence window) and, leading to a maximum number of counts $N_{\max} = T/\Delta w$ (e.g., 2×10^8 counts). From the Laplacian definition we find the counting probabilities:

$$P_3 = \frac{N_3}{N_{\max}} = N_3 \cdot \frac{\Delta w_c}{T};\ P_4 = \frac{N_4}{N_{\max}} = N_4 \cdot \frac{\Delta w_c}{T};\ P_c = \frac{N_c}{N_{\max}} = N_c \cdot \frac{\Delta w_c}{T} \Rightarrow$$
$$\alpha = \frac{P_c}{P_3 \cdot P_4} = \frac{N_c}{N_3 \cdot N_4} N_{\max} = \frac{N_c}{N_3 \cdot N_4}\left(\frac{T}{\Delta w_c}\right). \tag{A3}$$

where N_c is the number of coincident clicks of detectors D_3 and D_4. As above we introduced the ratio $\alpha = P(D_3 \,\&\, D_4)/[(P(D_3) \times P(D_4)]$.

Classical light with constant intensity I_0 is split into two beams, each carrying half of the energy (equal probability beam splitter). The detector will thus register $I_0/2$ from the incident irradiance I_0. Let us assume, the probability of a "click" being proportional to the light intensity (for small intensity) and to the sampling time Δt of the detector:

$$P_i = \eta_i \cdot I \cdot \Delta t,\ i = 3,\ 4. \tag{A4}$$

The probability of the two detectors clicking coincidentally is then given by

$$P_c = \eta_3\eta_4 \cdot I^2 \cdot (\Delta t)^2 \ \Rightarrow \alpha = \frac{P_c}{P_3 P_4} = \frac{\eta_3\eta_4 \cdot I^2 \cdot (\Delta t)^2.}{\eta_3 \cdot I \cdot \Delta t \cdot \eta_4 \cdot I \cdot \Delta t} = 1. \tag{A5}$$

Appendix A.4. Single Photons Interacting with Beam Splitters

Experimentally one finds (Figure 2/Section 3) that the probability of coincident clicks from detector D_3 and D_4, $P(D_3 \,\&\, D_4)$, vanishes and thus the substates are incompatible [8]. Quantum reasoning is now used for an explanation of this result.

Appendix A.4.1. Preparation of the Quantum States

The physical system of a single photon populating the output mode of a beam splitter can be described by a two-dimensional Hilbert space with the natural basis of a photon at beam splitter mode ③ and non at the mode ④ (see Figure A4): $|1_3\rangle = |1_3, 0_4\rangle$ and vice versa: $|1_4\rangle = |0_3, 1_4\rangle$. For the transformation between input and output we have to take into account two different possibilities, transmission and reflection, each with a particular probability and phase. This gives the recipe for preparing the output state.

The impact of a symmetrical beam splitter is described by one probability coefficient for the reflection, $r = \sqrt{p_r} q(\varphi_r)$, and another for the transmission, $t = \sqrt{p_t} q(\varphi_t)$, with $p_{r,t}$ as real probabilities and the phase factors $q(\phi)$ taking into account any phase jumps due to reflection or transmission. The input state is transformed by the beam splitter into a superposition of the substates {single photon state at beam splitter mode ③)} and {single photon state at beam splitter mode ④)}

$$|\psi_{\text{in}}\rangle \rightarrow |\psi_{\text{out}}\rangle = (\sqrt{p_r} q(\varphi_r) + \sqrt{p_t} q(\varphi_t))|1_3\rangle + (\sqrt{p_t} q(\varphi_t) + \sqrt{p_r} q(\varphi_r))|1_4\rangle. \tag{A6}$$

Checking for orthonormality fixes the meaning of p_r and p_t as probabilities and the values of the phase jumps. With $q(\phi_r - \phi_r) = q(\phi_t - \phi_t) = q(0) = 1$ we get $\phi_r - \phi_t = \pi/2$ [8]. Here, we use $\phi_r = \pi/2$ and $\phi_t = 0$. Inserting further conditions of an equal probability beam splitter ($p_r = p_t = 0.5$) we get the quantum states of the beam splitter experiment (see Table A2).

Figure A5 illustrates this transformation for the single photon incident in mode ①. The matrix $\hat{B}$ symbolizes the transformation by the beam splitter.

Table A2. Preparation of the quantum state at the bare beam splitter. See text and Figure A4 for details.

Operation	System State
Input: single photon state at the input mode ①	$\lvert\psi_{\text{in}}\rangle = \lvert 1_1, 0_2\rangle = \lvert 1_1\rangle$
Output: a superposition of the output modes ③ and ④	$\lvert\psi_{\text{out}}\rangle = \frac{1}{\sqrt{2}}(q(\pi/2)\lvert 1_3\rangle + \lvert 1_4\rangle)$

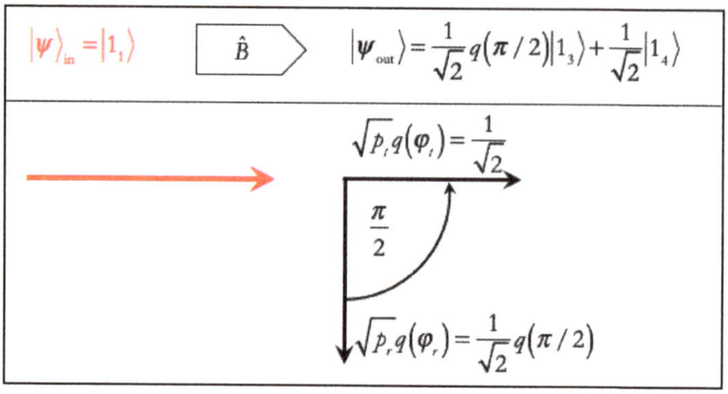

Figure A5. Pointer representation of output state superposition $|\psi_{out}\rangle$; the transformation $|\psi_{in}\rangle \rightarrow |\psi_{out}\rangle$ is mediated by the matrix $\hat{B}$.

Appendix A.4.2. Detection

The probabilities to find the photon in the output substates $|1_3\rangle$ or $|1_4\rangle$ are given by [59]:

$$\begin{aligned} P(|1_3\rangle) &= |\langle 1_3|\psi_{out}\rangle|^2 = \tfrac{1}{2}\left|q(-\tfrac{\pi}{2})\langle 1_3|1_3\rangle + \langle 1_3|1_4\rangle\right|^2 = \tfrac{1}{2}\left|q(-\tfrac{\pi}{2})\right|^2 = \tfrac{1}{2}, \\ P(|1_4\rangle) &= |\langle 1_4|\psi_{out}\rangle|^2 = \tfrac{1}{2}\left|q(-\tfrac{\pi}{2})\langle 1_4|1_3\rangle + \langle 1_4|1_4\rangle\right|^2 = \tfrac{1}{2}. \end{aligned} \quad \text{(A7)}$$

The interpretation is straightforward: one finds $P(|1_3\rangle) + P(|1_4\rangle) = 1$; the input photon will be detected from D_3 or D_4. This explains why coincident counts are not to be expected. This result, a lack of coincident counts, given by $P(D_3 \& D_4) = 0$, is shown in Figure 2b. The result indicates that the source produces single-photon states.

Appendix A.5. Single Photons Interacting with a Michelson Interferometer

In Equation (A7) a phase difference of $\pi/2$ is noticeable between the two superposed states $|1_3\rangle$ and $|1_4\rangle$. A phase difference between superposed substates will lead to interference fringes if one removes the spatial/temporal separation of the substates and detecting both states simultaneously superposed on one detector. Two further mirrors help.

The result is shown in Figure 3: Depending on the position of the mirror M_1 one gets interference fringes. To minimize noise, we measured the number N_{G2} of coincidence clicks of detector D_2 and a single photon trigger detector D_G, thus ensuring that only a single photon state is incident on the beam splitter (D_G is not shown in Figure 3). The visibility of the interference pattern is convincingly high. Inserting experimental data, for the visibility we get

$$V = \frac{N_{G2}(\max) - N_{G2}(\min)}{N_{G2}(\max) + N_{G2}(\min)} = \frac{3982}{4234} = 0.94. \quad \text{(A8)}$$

Appendix A.5.1. Preparation of the Quantum States

The operator procedure demonstrated in Equation (A6) is used here too. Again, the starting point is a single photon state at mode ① of the beam splitter. The impact of the beam splitting is thought to be the same as before. We thus can write down the total transformation chain for deriving the output state (see Table A3):

Table A3. Preparation of the quantum state at the interferometer.

Operation	System State	
Input	$\vert\psi_{\text{in}}\rangle = \vert 1_1\rangle$	
Beam splitting 1-step	$\vert\psi_{\text{out}/1}\rangle = \frac{1}{\sqrt{2}}\left(q\left(\frac{\pi}{2}\right)\vert 1_3\rangle + \vert 1_4\rangle\right)$	
Phase shift ϕ_M due to mirror 1 and mirror 2	$\vert\psi_{\text{out}/2}\rangle = \frac{1}{\sqrt{2}}\left(q\left(\frac{\pi}{2}\right)q(\varphi_{M2})\vert 1_3\rangle + q(\varphi_{M1})\vert 1_4\rangle\right)$	
Beam splitting 2-step	$\vert\psi_{\text{out}}\rangle = \frac{1}{2}\begin{bmatrix} (q(\pi)q(\varphi_{M2}) + q(\varphi_{M1}))\vert 1_1\rangle + \\ q\left(\frac{\pi}{2}\right)(q(\varphi_{M2}) + q(\varphi_{M1}))\vert 1_2\rangle \end{bmatrix}$	(A9)

Appendix A.5.2. Detection

Again we find the probability to detect photons at mode ① and mode ② (see Figure A6) using Born's rule,

$$\begin{aligned} P(\text{mode }(2)) = \left|\langle 1_2|\psi_{\text{out}}\rangle\right|^2 = & \frac{1}{4}\left| \begin{matrix} (q(\pi)q(\varphi_3)+q(\varphi_4))\underbrace{\langle 1_2|1_1\rangle}_{0} + \\ q\left(\frac{\pi}{2}\right)(q(\varphi_3)+q(\varphi_4))\underbrace{\langle 1_2|1_2\rangle}_{1} \end{matrix} \right|^2 = \frac{1}{4}[(q(-\varphi_3)+q(-\varphi_4))][(q(\varphi_3)+q(\varphi_4))] \\ & = \frac{1}{2}(1+\cos(\varphi_3-\varphi_4)) = \cos^2\left(\frac{\varphi_3-\varphi_4}{2}\right), \\ P(\text{mode }(1)) = \left|\langle 1_1|\psi_{\text{out}}\rangle\right|^2 = & \ \ldots = \sin^2\left(\frac{\varphi_3-\varphi_4}{2}\right). \end{aligned} \tag{A10}$$

We assume the interferometer is adjusted so that the state evolution between beam splitter and the mirrors leads to $\omega \times d/c = 2\pi$ (see Equation (A2)). The displacement of mirror 1 adds a phase $\Delta\phi$, thus $\phi_4 = 2\pi + \Delta\phi$:

This results expectedly gives the quantum interference pattern from Figure 3b. Quantum interference means, interference without waves. The experiment demonstrates transparent experimental evidence of the phase sensitivity of the probability amplitudes. The total probability, summed over the output modes equals one, as it should be. The pointer representation of Equation (A9) (Figure A6) illustrates this result: Adding the pointers gives the total quantum state, calculating the square of the pointer length gives the probability.

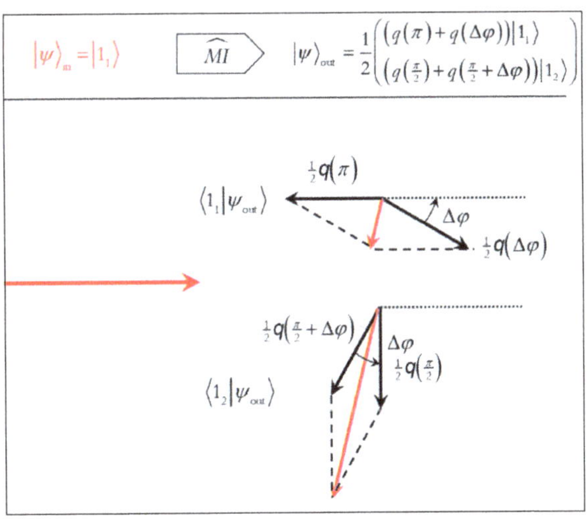

Figure A6. Pointer representation of single photon interference from a Michelson interferometer the transformation $|\psi_{\text{in}}\rangle \rightarrow |\psi_{\text{out}}\rangle$ is mediated by the matrix $\hat{M}i$.

References

1. Glauber, R. Coherent and incoherent states of the radiation field. *Phys. Rev.* **1963**, *131*, 2766–2788. [CrossRef]
2. Koopman, L.; Brouwer, N.; Heck, A.; Buma, W.J. Remedial mathematics for quantum chemistry. *J. Chem. Educ.* **2008**, *85*, 1233–1236. [CrossRef]
3. Müller, R.; Wiesner, H. Teaching quantum mechanics on an introductory level. *Am. J. Phys.* **2002**, *70*, 200–209. [CrossRef]
4. Dehlinger, D.; Mitchell, M.W. Entangled photon apparatus for the undergraduate laboratory. *Am. J. Phys.* **2002**, *70*, 898–901. [CrossRef]
5. Dehlinger, D.; Michtell, M.W. Entangled photons, nonlocality, and Bell inequalities in the undergraduate laboratory. *Am. J. Phys.* **2002**, *70*, 903–910. [CrossRef]
6. Galvez, E.J.; Holbrow, C.H.; Pysher, M.J.; Martin, J.W.; Courtemanche, N.; Heilig, L.; Spencer, J. Interference with correlated photons: Five quantum mechanics experiments for undergraduates. *Am. J. Phys.* **2005**, *73*, 127–140. [CrossRef]
7. Thorn, J.J.; Neel, M.S.; Donato, V.W.; Bergreen, G.S.; Davies, R.E.; Beck, M. Observing the quantum behaviour of light in an undergraduate laboratory. *Am. J. Phys.* **2004**, *72*, 1210–1219. [CrossRef]
8. Scholz, R.; Wessnigk, S.; Weber, K.-A. A classical to quantum transition via key experiments. *Eur. J. Phys.* **2020**, *41*, 055304. [CrossRef]
9. Kuhlmann, M. Interpretationen der Quantenfeldtheorie. In *Philosophie der Physik.*; Elsfeld, M., Ed.; Suhrkamp: Berlin, Germany, 2012.
10. Bitzenbauer, P.; Meyn, J.P. A new teaching concept on quantum physics in secondary schools. *Phys. Educ.* **2020**, *55*, 055031. [CrossRef]
11. Krijtenburg-Lewerissa, K.; Pol, H.J.; Brinkman, A.; van Joolingen, W.R. Insights into teaching quantum mechanics in secondary and lower undergraduate education. *Phys. Rev. ST Phys. Educ. Res.* **2017**, *13*, 010109. [CrossRef]
12. Kidd, R.; Ardini, J.; Anton, A. Evolution of the modern photon. *Am. J. Phys.* **1989**, *57*, 27–35. [CrossRef]
13. Muthukrishan, A.; Scully, O.M.; Zubairy, M.S. The concept of the photon—Revisited. In *The Nature of Light*; Roychoudhuri, C., Kracklauer, A.F., Creath, K., Eds.; CRC Press: Boca Raton, FL, USA, 2008; Chapter 4. [CrossRef]
14. Hong, C.K.; Ou, Z.Y.; Mandel, L. Measurement of subpicosecond time intervals between two photons by interference. *Phys. Rev. Lett.* **1987**, *59*, 2044–2046. [CrossRef] [PubMed]
15. Quantum Flagship. The future is quantum. Available online: https://qt.eu (accessed on 1 September 2022).
16. Masshadi, A.; Woolnough, B. Insights into students' understanding of quantum physics: Visualizing quantum entities. *Eur. J. Phys.* **1999**, *20*, 511–516. [CrossRef]
17. Michelini, M.G.; Ragazzon, R.; Santi, L.; Stefanel, A. Proposal for quantum physics in secondary school. *Phys. Educ.* **2000**, *35*, 406–410. [CrossRef]
18. Escalada, L.T.; Rebello, N.S.; Zollman, D.A. Student explorations of quantum effects in LEDs and luminescent devices. *Phys. Teach.* **2004**, *42*, 173–179. [CrossRef]
19. Hoekzema, D.; van den Berg, E.; Schooten, G.; van Dijk, L. The particle/wave-in-a-boxmodel in Dutch secondary schools. *Phys. Educ.* **2007**, *42*, 391–398. [CrossRef]
20. Lautesse, P.; Vila Valls, A.; Ferlin, F.; Héraud, J.L.; Chabot, H. Teaching quantum physics in upper secondary school in France. *Sci. Educ.* **2015**, *24*, 937–955. [CrossRef]
21. Henriksen, E.K.; Bungum, B.; Angell, C.; Tellefsen, C.W.; Fragat, T.; Vetleseter Bøe, M. Relativity, quantum physics and philosophy in the upper secondary curriculum: Challenges, opportunities and proposed approaches. *Phys. Educ.* **2014**, *49*, 678–684. [CrossRef]
22. Karakostas, V.; Hadzidaki, P. Realism vs. constructivism in contemporary physics: The impact of the debate on the understanding of quantum theory and its instructional process. *Sci. Educ.* **2005**, *14*, 607–629. [CrossRef]
23. Kalkanis, G.; Hadzidaki, P.; Stavrou, D. An instructional model for a radical conceptual change towards quantum mechanics concepts. *Sci. Educ.* **2003**, *87*, 257. [CrossRef]
24. Chi, M.T.H. Two kinds and Four sub-types of misconceived knowledge, ways to change it, and the learning outcomes. In *International Handbook of Research on Conceptual Change*; Vosniadou, S., Ed.; Routledge: New York, NY, USA; London, UK, 2013; Chapter 3. Available online: https://www.routledgehandbooks.com/doi/10.4324/9780203154472.ch3 (accessed on 1 September 2022).
25. Mortimer, E.F. Conceptual change or conceptual profile change? *Sci. Educ.* **1995**, *4*, 267–285. [CrossRef]
26. Waitzmann, M. Wirkung Eines Quantenoptischen Arguments auf Quantenphysikalische Argumentation. Ph.D. Thesis, Leibniz University Hannover, Hanover, Germany, 2022.
27. Posner, G.J.; Strike, K.A.; Hewson, P.W.; Gertzog, W.A. Accommodation of a scientific conception: Toward a theory of conceptual change. *Sc. Edu.* **1982**, *66*, 211–227. [CrossRef]
28. diSessa, A.A. Toward an epistemology of physics. *Cogn. Instr.* **1993**, *10*, 165–255. [CrossRef]
29. diSessa, A.A. Conceptual change in a microcosm: Comparative learning analysis of a learning event. *Hum. Dev.* **2017**, *60*, 1–37. [CrossRef]
30. McNeill, K.L.; Krajcik, J. Inquiry and scientific explanations: Helping students use evidence and reasoning. In *Science as Inquiry in the Secondary Setting*; Luft, J., Bell, R., Gess-Newsome, J., Eds.; NSTA Press: Arlington, VA, USA, 2008; Chapter 11. Available online: https://dpi.wi.gov/sites/default/files/imce/science/Inquiry_and_Scientific_Explanations_McNeill-Krajcik.pdf (accessed on 1 September 2022).

31. Chiara, M.; Giuntini, R.; Greechie, R. *Reasoning in Quantum Theory. Sharp and Unsharp Quantum Logics*; Springer Science + Business Media: Dordrecht, The Netherlands, 2004. [CrossRef]
32. Küblbeck, J.; Müller, R. *Die Wesenszuege der Quantenphysik: Modelle, Bilder und Experimente*; Aulis Verlag Duebne GmbH & Co KG: Köln, 2003. Available online: https://docplayer.org/40006159-Die-wesenszuege-der-quantenphysik-modelle-bilder-und-experimente.html (accessed on 1 September 2022).
33. Friebe, C.; Kuhlmann, M.; Lyre, H.; Näger, P.; Passon, O.; Stöckler, M. *Philosophie der Quantenphysik*; Springer Spektrum: Beilin/Heidelberg, Germany, 2015. [CrossRef]
34. Bitzenbauer, P. Effect of an introductory quantum physics course using experiments with heralded photons on preuniversity students' conceptions about quantum physics. *Phys. Rev. ST Phys. Educ. Res.* **2021**, *17*, 020103. [CrossRef]
35. Aguayo, R.; Simms, G.; Siegel, P.B. Throwing nature's dice. *Am. J. Phys.* **1996**, *64*, 752–758. [CrossRef]
36. Bronner, P.; Strunz, A.; Silberhorn, C.; Meyn, J.P. Demonstrating quantum random with single photons. *Eur. J. Phys.* **2009**, *30*, 1189–1200. [CrossRef]
37. Ireson, G. The quantum understanding of pre-university physics students. *Phys. Educ.* **2000**, *35*, 15–21. [CrossRef]
38. Pospiech, G.; Schöne, M. Quantenphysik in Schule und Hochschule. *PhyDid B—Didaktik der Physik—Beiträge zur DPG-Frühjahrstagung*. 2012. *Mainz*, 20.01. Available online: https://ojs.dpg-physik.de/index.php/phydid-b/article/view/392 (accessed on 1 September 2022).
39. Zollman, D.A.; Rebello, N.S.; Hogg, K. Quantum mechanics for everyone: Hands-on activities integrated with technology. *Am. J. Phys.* **2002**, *70*, 252–259. [CrossRef]
40. Bronner, P. *Quantenoptische Experimente als Grundlage eines Curriculums zur Quantenphysik des Photons*; Logos Verlag: Berlin, Germany, 2010. Available online: https://d-nb.info/100496515X (accessed on 1 September 2022).
41. Häußler, P. *Physikalische Bildung: Eine curriculare Delphi-Studie. Teil 1., IPN-Arbeitsbericht 41*; IPN: Kiel, Germany, 1980.
42. Burkard, U.E. *Quantenphysik in der Schule: Bestandsaufnahme, Perspektiven und Weiterentwicklungsmöglichkeiten durch die Implementierung Eines Medienservers*; Logos-Verlag: Berlin, Germany, 2009. Available online: https://www.fachportal-paedagogik.de/literatur/vollanzeige.html?FId=1057488 (accessed on 1 September 2022).
43. Weber, K.-A.; Friege, G.; Scholz, R. Quantenphysik in der Schule—Was benötigen Lehrkräfte?—Ergebnisse einer Delphi-Studie. *Zeit. Didaktik Natur.—ZfDN* **2020**, *26*, 173–190. [CrossRef]
44. Bitzenbauer, P.; Meyn, J.P. Fostering students' conception about the quantum world—Results of an interview study. *Prog. Sc. Ed.* **2021**, *4*, 40–51.
45. Marshman, E.; Singh, C. Framework for understanding the patterns of student difficulties in quantum mechanics. *Phys. Rev. ST Phys. Educ. Res.* **2015**, *11*, 02011. [CrossRef]
46. Ubben, M.; Heusler, S. Gestalt and functionality as independent dimensions of mental models in science. *Res. Sc. Educ.* **2019**, *51*, 1349–1363. [CrossRef]
47. Kral, A.; Theis, C.; Schorn, B.; Heinke, H. Praxistauglicher Einstieg in die Quantenphysik mit Realexperimenten. In *Implementation fachdidaktischer Innovation im Spiegel von Forschung und Praxis. Gesellschaft für Didaktik der Chemie und Physik Jahrestagung in Zürich 2016*; Maurer, C., Ed. Universität Regensburg: Regensburg, Germany, 2017; pp. 400–404.
48. Niebert, K.; Gropengießer, H. Leitfadengestützte Interviews. In *Methoden in der Naturwissenschaftsdidaktischen Forschung*; Krüger, D., Parchmann, I., Schecker, H., Eds.; Springer Spektrum: Heidelberg/Berlin, Germany, 2014; pp. 121–132. [CrossRef]
49. Ayene, M.; Kriek, J.; Damtie, B. Wave-particle duality and uncertainty principle: Phenomenographic categories of description of tertiary physics students' depictions. *Phys. Rev. ST Phys. Educ. Res.* **2011**, *7*, 020113. [CrossRef]
50. Kuckartz, U. *Qualitative Inhaltsanalyse. Methoden, Praxis, Computerunterstützung*; Beltz Juventa: Weinheim, Germany, 2016.
51. Schreier, M. Varianten qualitativer Inhaltsanalyse: Ein Wegweiser im Dickicht der Begrifflichkeiten [59 Absätze]. *Forum Qual. Soc. Res.* **2014**, *15*, 18. [CrossRef]
52. Döring, N.; Bortz, J. *Forschungsmethoden und Evaluation in den Sozial- und Humanwissenschaften*; Springer: Berlin/Heidelberg, Germany, 2016. [CrossRef]
53. Mannheim, E.G.; Helmes, A.; Weis, J. Tanztherapie in der stationären Onkologischen Rehabilitation. *Forsch. Komplement.* **2013**, *20*, 33–41. [CrossRef]
54. Greenstein, G.; Zajonc, A.G. *The Quantum Challenge*; Jones and Barlett Publishers, Inc.: Sudbury, MA, USA, 1997. Available online: https://xn--webducation-dbb.com/the-quantum-challenge-pdf/ (accessed on 1 September 2022).
55. Messiah, A. *Quantum Mechanics*; Dover Publication: New York, NY, USA, 1999.
56. Filk, T. *Quantenmechanik (Nicht nur) für Lehramtsstudierende*; Springer Spektrum: Berlin/Heidelberg, Germany, 2019.
57. Scully, M.O.; Zubairy, M.S. *Quantum Optics*; Cambridge University Press: Cambridge, UK, 1997. [CrossRef]
58. Feynman, R.P.; Zee, A. *QED: The Strange Theory of Light and Matter*; Princeton University Press: Princeton, NJ, USA, 2006.
59. Born, M. Zur Quantenmechanik der Stoßvorgänge. *Z. Für Phys.* **1926**, *37*, 863–867. [CrossRef]
60. Loudon, R. *The Quantum Theory of Light*; Oxford University Press: Oxford, UK, 2000. Available online: http://rplab.ru/~as/2000%20-%20R.Loudon%20-%20The%20Quantum%20Theory%20of%20Light%20-%203rd%20ed%20Oxford%20Science%20Publications.pdf (accessed on 1 September 2022).

physics MDPI

Article

An Approach to Quantum Physics Teaching through Analog Experiments

Stefan Aehle [1,*], Philipp Scheiger [1,2] and Holger Cartarius [1]

[1] Working Group Teaching Methodology in Physics and Astronomy, Friedrich-Schiller-University Jena, 07743 Jena, Germany
[2] Physics Education Research, 5th Institute of Physics, University of Stuttgart, 70550 Stuttgart, Germany
* Correspondence: stefan.aehle@uni-jena.de

Abstract: With quantum physics being a particularly difficult subject to teach because of its contextual distance from everyday life, the need for multiperspective teaching material arises. Quantum physics education aims at exploring these methods but often lacks physical models and haptic components. In this paper, we provide two analog models and corresponding teaching concepts that present analogies to quantum phenomena for implementation in secondary school and university classrooms: While the first model focuses on the polarization of single photons and the deduction of reasoning tools for elementary comprehension of quantum theory, the second model investigates analog Hardy experiments as an alternative to Bell's theorem. We show how working with physical models to compare classical and quantum perspectives has proven helpful for novice learners to grasp the abstract nature of quantum experiments and discuss our findings as an addition to existing quantum physics teaching concepts.

Keywords: physics education; quantum theory; quantum technology education; history of physics in physics education; philosophy of physics in physics education

1. Introduction

Due to its inherently unintuitive nature, quantum theory has been met with disbelief and partly even resentment since its conception in the early twentieth century. Accordingly, most physicists have been struggling with overcoming the established concepts of the deterministic and local worldview—seasoned experts maybe even more so than novice learners. One of the most known examples is the classic Einstein–Podolsky–Rosen (EPR) paradox [1]. However, even for younger minds, it is often challenging to accept the results provided by modern quantum experiments and the implications that arise thereof [2], even though modern views on quantum physics that emphasize its comprehensibility and help with the interpretation are now widespread and have found their way into standard textbooks [3–5]. For that reason, educators have since been trying to enrich their teaching with thought experiments, mental models, and analogies [6] to allow their audience to develop an understanding from multiple perspectives. Luckily, with the growing importance of quantum physics for modern society and industry [7], research in the field of quantum physics education has progressed to explore a variety of options bringing advanced topics and methods into secondary school and university classrooms: quantum computing [8], quantum randomness [9,10], quantum sensing, and new concepts applying quantum technologies to a changing educational framework [11–14]. Recent empirical studies play an equally important role in evaluating the overall progress of the field [7,15,16]. In addition, the growing field of edutainment, combining modern multimedia technologies such as augmented and virtual reality with learning experiences, has been explored for quantum education [17–19].

Education psychology attributes validity to all of these approaches by emphasizing the importance of multisensory learning materials: cognitive-affective theory of learning with media (CATLM) suggests consideration of not just auditory and visual sources but tactile and haptic elements as well [20–23], and object-based learning (OBL) claims "that haptic interaction with real tangible objects can serve important roles in the learning process and encourages students to link these experiences to abstract ideas and concepts" [20]—an idea that extends through education ever since physical models, exemplary specimens, and other display items decorated walls, shelves, and school supply rooms waiting for their deployment in a lesson period.

On the same level, especially considering physics education, the use of analogies becomes an undeniably important factor since most physics teaching is based on the simplification and idealization of complex processes. The use of models and analogies appears inevitable, bridging the gap between empirical and theoretical entities through "familiarity" [20]. In that way, quantum physics teaching often lacks the ability to build on familiarity just because the subject matter does not allow comparison to the macroscopic world. Nevertheless, we are aiming to fill the gap of missing analog models and add to the already existing representations of quantum phenomena.

In what follows, we present two applications of analog experiments of quantum phenomena for teaching scenarios. The first application addresses the polarization of single photons in Section 2 and showcases a model based on near-field communication (NFC) technology, Arduino microcontrollers, and 3-dimensional (3D) printing. The corresponding teaching concept focuses on a multiperspective approach, comparing classical and quantum explanations of polarization through Section 2.1 while following the study of Müller and Küblbeck [24], who laid most accepted foundations for modern quantum physics teaching. All technical aspects of the polarization analogy modules are mapped out in Section 2.2 before their implementation into quantum physics teaching is featured in Section 2.3. The second application (Section 3) employs a different analog architecture based on mechanical rather than electronic components: combining clear formalism of quantum mechanics with a local representation, we propose a model to disprove the local hidden variable theory, following the argumentation of Lucien Hardy [25]. Section 3.1. summarizes these considerations about nonlocality regarding entangled spin-states of electrons in a Stern–Gerlach experiment [6]. while Section 3.2. motivates its usage in a (higher) education context. Conclusions are drawn in Section 4.

2. Analog Experiment: Polarization Modules

2.1. Comparison of Quantum and Classical Polarization from a Teaching Perspective

Talking about polarization in a secondary school classroom is usually conducted within a unit about mechanical oscillations and waves and again at an advanced stage when talking about the properties of light [26]. Especially the latter leads to a dilemma, which is conventionally used as a stepping stone into the field of modern physics and quantum mechanics: What *is* light? At that point, students have seen light behave as a ray, a wave, or a particle, depending on the examined experiment and context. They learn that light may exhibit all of these characteristics, but the individual photon is, in fact, something entirely new: an object called a *quantum* object. The common narrative then explains the quantum object to be governed by a set of peculiar, unintuitive rules, fully unfathomable to the human mind. However, through experimentation, a set of "reasoning tools" [27] can be found, which then help create a conceptual approach to quantum physics. The following Sections compare quantum and classical explanations of polarization and show how the analog model mediates between both perspectives. The quantum aspects are only discussed on the level of clearly defined photon number states, i.e., we do not take states with undefined photon numbers into account. While there are reasons, even in a school context, to discuss the existence of quantum states of light, in which the photon number is not well defined [28], this is not required for the educational purpose of this work.

2.1.1. Classical Perspective

Most curricula introduce light as a wave by referring to mechanical transverse waves. As such, it shares its most common features, such as diffraction behaviors and the ability to create interference and be polarized. A linear polarizer filters out any part of the wave that does not match the polarization axis of the filter, thus diminishing light intensity. The remainder is transmitted and oriented along that axis—hence polarized. What happens when this transmitted, polarized portion meets a second polarizer is expressed by Malus' law of the light intensity [3],

$$I = I_0 \cdot \cos^2(\alpha). \tag{1}$$

Light as an electromagnetic wave is thought to pass a linear polarizer *partially*: Its oscillating vector (result of electric and magnetic field vector, respectively) is split into components vertical and horizontal to the polarization axis of the filter—one part being transmitted and the other being absorbed. The angle α between the first and second polarization affects how much of the initial intensity, I_0, will be measured behind the second filter.

The cosine dependency of Malus' law (1) can be tested qualitatively and even quantitatively in most classroom settings. A light meter is sufficient.

2.1.2. Quantum Perspective

One conclusion that is carried over from a physics curriculum's explanation of the photoelectric effect and the black body radiation problem to this polarization case is the quantization of photon energy. Thinking of light as discrete packages of energy called photons leads to the idea that light cannot pass a polarizer just *partially*. The individual photon may only be transmitted *entirely* or not at all [29]. A detector behind a polarizing filter will—within a certain time interval—either detect the photon or not. It is impossible to make a statement about the behavior of a single photon until we measure its presence (or absence) at the output of the experiment. In other words: quantum objects are inherently governed by randomness. Students may arrive at this profound conclusion with the help of a teacher guiding their train of thought or possibly even on their own. Either way, abandoning classical determinism is key to engaging with quantum phenomena.

Nevertheless, experiments on quantum systems show that statistical predictions about results are possible. For polarization problems, this means there is a certain probability for the photon to be transmitted, to be in a transmission state, $|\Psi_T\rangle$, and pass the filter or to be absorbed and be in an absorption state, $|\Psi_A\rangle$, when it interacts with the filter. To calculate these probabilities (see Figure 1), we define the photon's polarization state before the filter,

$$|\Psi_P\rangle = \cos\theta_1|\leftrightarrow\rangle + \sin\theta_1|\updownarrow\rangle, \tag{2}$$

as a combination of vertical, $|\updownarrow\rangle$, and horizontal, $|\leftrightarrow\rangle$, vectors in a reference basis. θ_1 is the angle the polarization state is rotated from $|\leftrightarrow\rangle$. Let us further define a measurement basis representing the second polarizer. This second polarizer will either transmit or absorb—in other words, *measure*—the photon. It is rotated by an angle θ_2 to the reference basis and holds the transmission and absorption states, $|\Psi_T\rangle$ and $|\Psi_A\rangle$, which we define in relation to the reference vectors:

$$|\Psi_T\rangle = -\sin\theta_2|\leftrightarrow\rangle + \cos\theta_2|\updownarrow\rangle, \tag{3}$$

$$|\Psi_A\rangle = \cos\theta_2|\leftrightarrow\rangle + \sin\theta_2|\updownarrow\rangle. \tag{4}$$

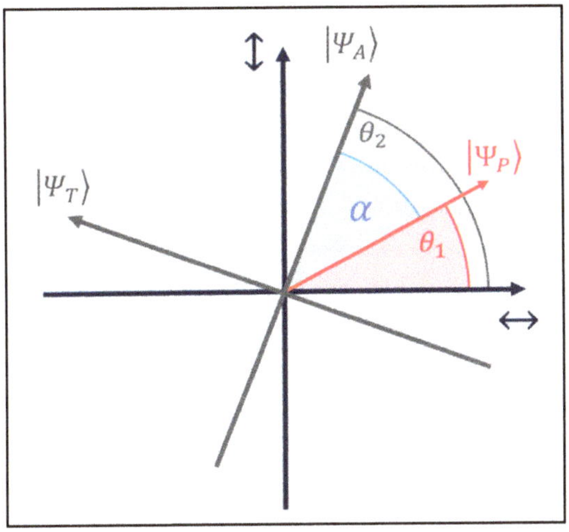

Figure 1. Visualization of polarization states of photons at a polarizer. To determine transmission and absorption probabilities, the polarization state $|\Psi_P\rangle$ is projected onto $|\Psi_T\rangle$ and $|\Psi_A\rangle$, respectively. See text for details.

One is now ready to calculate the probabilities, P_T and P_A, respectively, of transmitting or absorbing the photon at the second filter by projecting the polarization state onto the measurement basis, which leads to:

$$P_T = |\langle \Psi_T | \Psi_P \rangle|^2 = \cos^2(\theta_1 - \theta_2), \tag{5}$$

$$P_A = |\langle \Psi_A | \Psi_P \rangle|^2 = \sin^2(\theta_1 - \theta_2). \tag{6}$$

As it turns out, the same cosine dependency of the angle $\alpha = (\theta_1 - \theta_2)$ is found between the two polarizers, as Malus' law (1) states for classical light intensity: what defines the transmission probability of the single photon in a quantum state of a well-defined photon number manifests itself as statistical evidence on a macroscopic level.

Using Dirac's bra–ket notation, < ... >, simplifies the quantum mechanical calculation and could, for that reason, be used even in secondary school classrooms. However, in the next Section we show that the understanding been derived from this calculation may also be reached experimentally using the analog model.

2.2. Polarization Analogy Module

After considering a number of different methods to implement the extraordinary phenomena of quantum physics into a physical model, a hybrid approach of mechanical and electronic elements was settled on: The current iteration of the project takes the shape of a black box that reads, writes, and redirects incoming NFC chips through utilizing servo motors and an Arduino microcontroller. NFC chips are able to store and carry information passively without the need for additional battery power. They are activated through a radio frequency identification (RFID) module controlled by the Arduino. An Arduino can connect and control several different input and output components, such as potentiometers, sensors, motors, displays, and buttons, at the same time. The combination of all these electronic parts with physical elements such as ducts, switches, and flaps, all within a suitable enclosure, guide the NFC chips within a physical model representing quantum measurements and results. We find this hybrid approach to be the most flexible and best-suited platform for creating analogies for a variety of real-world experiments on quantum systems. Figure 2 shows images of the current prototyping state of the model.

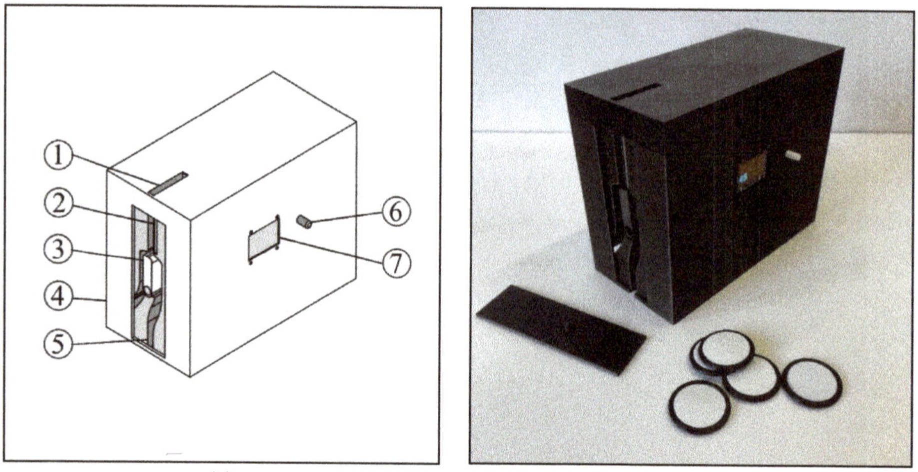

(a) (b)

Figure 2. Prototype of the polarization analogy module. (**a**) Isometric view: (1) entry slot for near-field communication (NFC) chips, (2) location of the embedded radio-frequency identification (RFID) module to interact with NFC chips, (3) hinged flap to guide chips after reading/writing, (4) back exit slot, (5) bottom exit slot, (6) control knob, (7) organic light-emitting diode (OLED) display. (**b**) Photo of the 3-dimensional (3D)-printed model and NFC chips: Values from 0 to 90 can be set by turning the control knob next to the display. This value will be programmed into the chips on their way through the system. The white coin-sized NFC chips are encapsulated in a black cover to improve operability by increasing their grip. A removable side panel reveals the inner paths.

Functionally, this model resembles the "behavior" of single photons at a polarizing filter. Though one cannot speak of what the photon actually does, one can conduct experiments and take measurements that lead to definite results and tangible interpretations. Specifically, one knows that the photon as a quantum object will either be fully transmitted or fully absorbed by the filter. Furthermore, we know that the probability of either of those events is dependent on the polarization state of the photon in relation to the polarization axis of the filter. These results are replicated by the model relating to the following analogy:

Single photons passing a polarizer are represented by individual NFC chips passing through the black box in the following, called the polarization module (Figure 2). Similar to photons, these chips will either be transmitted (leaving through the bottom exit slit) or absorbed (leaving through the back exit slit). Additionally, they are given a polarization state, corresponding to the polarization axis of the filter, that is written onto the chip and saved as an integer value. This value is displayed at the front of the module and can be set between 0 degrees and 90 degrees by turning the control knob. Adjusting this value is equivalent to rotating the orientation of a physical polarizer when light is passed through. Which path the chips take within the polarization module is determined by the Arduino microcontroller. First, it reads the polarization state of an incoming chip and compares this value with the setting on the display, i.e., the orientation of the polarization axis of the filter. Second, it calculates the difference between the two angles and the resulting transmission and absorption probability, respectively. Third, it uses a pseudorandom number with respect to that probability to choose the path in which the chip will be redirected.

Polarization phenomena may not be the prime example to emphasize the fundamental differences between classical and quantum physics (that may be the double slit experiment); however, it may serve as yet another case in which to derive and discuss these differences. As shown in the following, for the early stages, this can be done even without a deep understanding of the concept of interference (as is required for the double slit).

2.3. Polarization Analogy and Reasoning Tools

Comparing the classical and quantum perspectives of the same phenomena allow learners to identify key differences between the classical and the modern worldview. They are enabled to generalize their findings, define a set of rules—or rather reasoning tools—and make predictions about the outcome of experiments. So far, it has been explained what this comparison yields theoretically. The following describes what this multiple perspective approach, supported by physical analog models, looks like when practiced with a group of students entirely new to the field of quantum physics.

2.3.1. Conceptualization of a Students' Laboratory

The polarization analogy modules (Section 2.2) have been designed as a platform to be introduced in physics classrooms and student labs in a variety of different use cases. Their hardware, software, and lesson integration can be altered in any way to fit different preconditions, learner types, time frames, and classroom layouts. Even their contextual setting as "polarization filters" might, at some point, be changed to suit another narrative. Therefore, the scenario presented here, which describes a first learning setup, may be understood as a promising yet preliminary result. The concept at hand was developed with a certain one-size-fits-all mentality in mind, with the purpose of accommodating student groups at different learning stages as follows:

A visiting group of students at our laboratory is separated into small peer groups of two to three learners (11 students in total). Depending on their knowledge prior to the lesson, they receive a short theoretical introduction (in part, a revision of wave optics) to particle–wave duality of light, leading up to the question: What is a quantum object, and how does it behave? Afterward, the main goal of their 2-h-visit is to answer that question and to find out how photons behave as quantum objects. The peer groups are then asked to work at four stations simultaneously, switch after a certain amount of time, and discuss and compare their findings with the rest of the class at the end of the lesson. Three of the four stations are set up with a number of polarization analogy modules, while the fourth station allows the learners to experiment with actual polarizers and lasers on an optical bench. Throughout the entire lesson, students are guided by supervisors.

Thematically, the first two stations form a unit regarding the statistical behavior of quantum objects. Having arrays of two to three analog modules (representing two to three polarizers in a row) available at each table, students are tasked to drop NFC chips (photons) into their models and make predictions about each individual outcome. Station 1 consists of two of the modules introduced in Section 2.2. The first module *prepares* the state of the photon, writing a value onto the chips. The second module then *measures* the state of the photon by reading the value off of the chip. It is then to be determined what effect the angle α between the two polarization axes has (especially considering special case angles of 0 and 90 degrees). After adding a third module to the array (station 2), the experiment can be altered to check if the same findings remain true when a third polarization axis is added to the system. Meanwhile, the students take notes of the results, list occurrences of each possibility, and calculate percentages of transmission events with respect to the angle between two polarizers. They draw conclusions and make generalizations about what would affect the "behavior" of single photons and about how their choice of polarization angles would influence those results. Finally—though not as abstract and generalized, but in conversation with supervisors and peers—the learners are then aiming to summarize what Küblbeck and Müller called "Wesenszüge" [24] and Müller and Mishina translated as "reasoning tools" [27]: a set of rules allowing to axiomatically describe the essence of quantum physical phenomena.

- Rule 1—Statistical Behavior. Single events are not predictable; they are random. Only statistical predictions (for many repetitions) are possible in quantum physics.
- Rule 2—Unique Measurement Results. Even if quantum objects in a superposition state need not have a fixed value of the measured quantity, one always finds a unique result upon measurement.

Similarly, stations 3 and 4 are united by their subject matter. These stations are used to strengthen the understanding reached so far by linking the behavior of single photons to the statistical result observed with a classical light wave. At station 3, the findings of the first two stations are to be built upon by extending the array of polarizers by another pair, such that the students now examine the statistics of photons passing through four or five successive filters. Additionally, while the first and last filters are oriented 90 degrees apart from each other, the polarizers in between are to be adjusted to transmit as much light as possible. This task is then replicated at station four using classical light, actual polarizers, and an optical bench. The resulting intensity after the last polarizer is measured with a light meter and placed in relation to the original intensity of the laser. The ratio of these two intensities is then compared with the statistics observed in station 3. In addition, station 4 allows the students to test Malus' law (1) qualitatively.

The study at hand only includes student tasks to find Rules 1 and 2 of the reasoning tools. However, the original set comprises four Rules. Let us list here the remaining two Rules for completeness:

- Rule 3—Ability to Interfere. Interference occurs if there are two or more "paths" leading to the same experimental result. Even if these alternatives are mutually exclusive in classical physics, none of them will be "realized" in a classical sense.
- Rule 4—Complementarity. Exemplary formulations are: "which path information and interference pattern are mutually exclusive" or "quantum objects cannot be prepared for position and momentum simultaneously".

A reasonable discussion of interference should include interference patterns, which need some spatial resolution to be clearly observed. Naturally, such resolution cannot be implemented with the two-way output of the analog modules: the binary output information is not sufficient. However, the Arduino platform is highly flexible, and a multiple-output setup is possible. In the future, this will be used to extend the teaching concept with analog modules to complete the set of reasoning tools.

2.3.2. Results and Students' Evaluation

As mentioned before, so far, we have been able to test this teaching concept with a group of secondary school students. The participating group arrived in our laboratory without a prior introduction to quantum physics. However, in preparation for their visit, they had been reviewing topics of wave optics, such as interference, polarization of light, and even particle–wave duality. Though the sample size is too small to extrapolate actual empirical evidence, we were able to register a significant increase in content knowledge, which can be described as follows.

Primarily, for internal review, the students were asked to participate not just in the testing of the analog polarization modules but in a contextual pre-test and post-test designed to evaluate their general understanding of the subject matter (see Table 1). Before and after their laboratory visit, all of the 11 students answered a set of comprehension items, deciding whether they were true or false. The increase in understanding was expressed in the following items: "A photon at a polarization filter can be absorbed partially" (pre-test: 2 correct answers, post-test: 9); "The measurement of the state of a quantum object does not affect that state" (pre-test: 4 correct answers, post-test: 6); "Compared to classical experiments, the results of quantum experiments are governed by randomness" (pre-test: 8 correct answers, post-test: 11). Hence, in the second run after the lesson, the rate of correctly answered items was significantly higher. Surprisingly, one item did not follow that trend and triggered more incorrect answers in the second round: "Quantum phenomena can only be verified through thought experiments and analogies, not real experiments" (pre-test: 5 correct answers, post-test: 2). In the future, more attention to be paid to this result to avoid misconceptions from only focusing on analog experiments. One possible solution might lie in expanding this setup through an actual quantum experiment.

Table 1. Comparison of pre-test and post-test results. A total of 110 comprehension items—10 items per person—was answered by 11 students before and after their lesson to track comprehension. Correct answers increased from 51% to 68%.

	Pre-Test		Post-Test	
	Absolute	Percentage	Absolute	Percentage
Correct Answers	56/110	51%	75/110	68%
Incorrect Answers	54/110	49%	35/110	32%

Overall, the students were able to gain a deeper understanding of how quantum objects "behave" while being introduced to concepts that will help shape an easier transition into learning about quantum physics. We were able to convey the gist of Section 2.1.2. in a student-centered approach, implementing different perspectives and experiments—rather untypical for quantum physics teaching.

3. A Hidden Parameter and Hardy's Experiment

The nonlocal nature of quantum mechanics is hard to believe since it contradicts the experience we obtained in the classical world. This is, in particular, true for novice learners. When they first hear about entanglement, the concept of hidden parameters seems to be appealing. With the following second analog experiment, we address this point and introduce a possibility to disprove one (at first glance) attractive idea to avoid nonlocality. This is performed with a non-complicated choice of a (local) hidden parameter. It might seem to be quite critical to place so much emphasis on hidden parameters since this could lead to misunderstandings (which happens often enough in this context [30,31]). The use of hidden parameters could distract from the importance of nonlocality. This should be discussed in any teaching concept.

Analogous experiments that imitate entanglement already possess hidden parameters since they are based on classical objects. Certain material properties such as weight, shape, magnetism, or a digital number on a chip can be such hidden parameters. We see great potential for teaching with analog experiments to first illustrate and discuss relevant terms and concepts (such as reality and locality). Afterward, a quantum mechanical experiment can be thought through/played out with a hidden parameter until a contradiction to the real experiment occurs, and, thus, a conceptual change, showing that the first appealing concept of local hidden parameters is not a good choice, can be triggered. This, certainly does not clarify that nonlocality is the only required assumption for the correct prediction of quantum physics. However, it provides a relevant step in the discussion with learners.

Actually, Bell's inequalities are used to disprove a hidden parameter in the experiment [32]. However, we present here an experiment proposed by Lucien Hardy [25]. This experiment was chosen as soon as its hidden parameter variant is not too complicate to be reproduced with an analog experiment, and the chain of reasoning to contradict quantum physics with a hidden parameter is quite straightforward and, therefore, more accessible for students.

3.1. *Nonlocality for Two Particles without Inequalities for Almost All Entangled States*

The experiment starts with a system of two entangled particles, where α and β are two real constants with $\alpha^2 + \beta^2 = 1$:

$$|\Psi\rangle = \alpha|\uparrow\rangle_1|\uparrow\rangle_2 - \beta|\downarrow\rangle_1|\downarrow\rangle_2. \tag{7}$$

The arrows in this notation represent orientations of electron spins in the z-direction, which can be measured in a Stern–Gerlach experiment. The two particles are sent to two different researchers (Alice and Bob) with different Stern–Gerlach experiments. Alice and Bob can measure the spin orientation in the z-direction and two other directions tilted by the angles θ_1 and θ_2, where $\tan(\theta_1/2) = \sqrt{\alpha/\beta}$ and $\tan(\theta_2/2) = -(\alpha/\beta)^{3/2}$. These tilted measurements can be expressed in new bases. There is the skewed θ_1-basis with basis states "up", $|\nearrow\rangle$, and "down", $|\swarrow\rangle$, and the horizontal θ_2-basis with the basis states "right", $|\rightarrow\rangle$, and "left", $|\leftarrow\rangle$. The initial state can now, according to quantum theory, be expressed for four different measurement cases.

Alice and Bob both decide to measure the spin in θ_1-direction:

$$|\Psi\rangle = \mathrm{N}\Big(AB|\nearrow\rangle_1|\swarrow\rangle_2 + AB|\swarrow\rangle_1|\nearrow\rangle_2 + B^2|\swarrow\rangle_1|\swarrow\rangle_2\Big). \tag{8}$$

Alice measures the spin in θ_2 direction and Bob in θ_1-direction:

$$|\Psi\rangle = \mathrm{N}\Big(\Big(A - A^2A^*\Big)|\rightarrow\rangle_1|\nearrow\rangle_2 + B|\rightarrow\rangle_1|\swarrow\rangle_2 - A^2B|\leftarrow\rangle_1|\nearrow\rangle_2\Big). \tag{9}$$

Alice measures the spin in θ_1 direction and Bob in θ_2-direction:

$$|\Psi\rangle = \mathrm{N}\Big(\Big(A - A^2A^*\Big)|\nearrow\rangle_1|\rightarrow\rangle_2 + B|\swarrow\rangle_1|\rightarrow\rangle_2 - A^2B|\nearrow\rangle_1|\leftarrow\rangle_2\Big). \tag{10}$$

Alice and Bob both decide to measure the spin in θ_2-direction:

$$|\Psi\rangle = \mathrm{N}\Big(\Big(1 - |A|^4\Big)|\rightarrow\rangle_1|\rightarrow\rangle_2 + A^2A^*B|\leftarrow\rangle_1|\rightarrow\rangle_2 + A^2A^*B|\rightarrow\rangle_1|\leftarrow\rangle_2 - A^2B^2|\leftarrow\rangle_1|\leftarrow\rangle_2\Big). \tag{11}$$

Here, $N = \frac{1-|\alpha\beta|}{|\alpha|-|\beta|}$, $A = \frac{\sqrt{\alpha\beta}}{\sqrt{1-|\alpha\beta|}}$ and $B = \frac{|\alpha|-|\beta|}{\sqrt{1-|\alpha\beta|}}$ are prefactors.

Equation (11) clearly states that in some cases, Alice and Bob can both measure spin "left", $|\leftarrow\rangle_2$, in the horizontal, i.e., θ_2-basis. If the measurement result is predetermined by a hidden parameter, then in such an experiment, the two particles must be prepared in such a way that if Alice and Bob measure in the θ_2-basis, both get spin "left" as a result.

If just such a spin pair were underway, but Bob had chosen to measure in the oblique θ_1-base, his result would also be predetermined according to Equation (9), as Bob must measure spin "up" in the θ_1-base because there is no other possibility in which the predetermined result "left" for Alice is preserved. When Bob chooses the skewed θ_1-basis so late that his measurement cannot affect Alice's measurement according to the laws of realism and locality, then a hidden parameter λ, which is still unknown but is supposed to fully describe quantum mechanics, must account for the result of Equation (9). The same is true in case Bob remains in the θ_2-basis and Alice chooses the skewed θ_1-basis according to Equation (10). Here, the only possibility for Alice is to obtain the value "up" in the θ_1-basis.

Finally, if both chose the Stern–Gerlach experiment in the θ_1-direction, the two cases with a hidden parameter λ discussed above indicate that both Alice and Bob must measure spin "up" since no setup knows the orientation of the other. However, such a measurement result is forbidden according to quantum theory, cf. Equation (8). Here lies the contradiction, which can be checked experimentally.

3.2. Didactical Framework

Our experience shows that students can have difficulties even with this simplified chain of reasoning. In order to support this graphically and haptically, we offer an analog experiment according to the following structure. The hidden parameter is the magnetizability of balls, which pass through a box with two-way junctions (see Figure 3). Magnetizable balls are deflected, and the others are not. Two Stern–Gerlach experiments each can now be selected by Alice and Bob, each representing a measurement in the θ_1 or θ_2 direction. The setup can be built to reproduce all the cases discussed above for a $|\leftarrow\rangle_1|\leftarrow\rangle_2$ pair. Thus, the hidden parameter setup provides the theoretical result for a hidden parameter quantum physics, i.e., one possible local description of quantum physics.

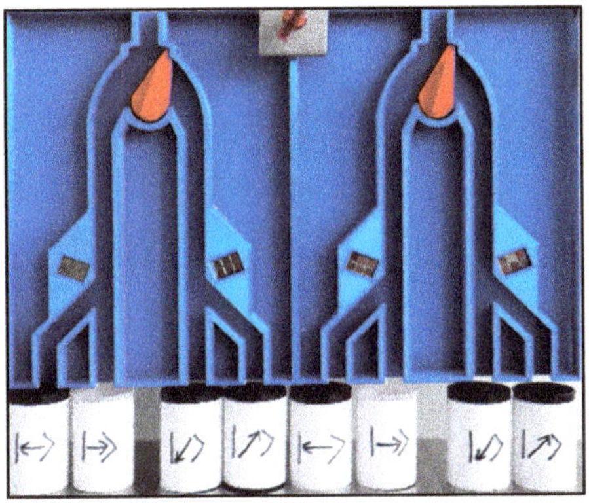

Figure 3. 3D-printed box with two-way connections. The left part represents Alice's measurement device; the right one, Bob's. The red handles each determine the orientation of the Stern–Gerlach experiment (the right path leads to a measurement in the θ_1 direction, the left in the θ_2 direction). Magnets at the bifurcation points guarantee the predetermined result for magnetizable balls. See text for details.

The setup is designed in a way that an electron pair with the hidden parameter "magnetizable" will reproduce the measurement result spin "left" for Alice and Bob when they decide to use the horizontal θ_2-orientated Stern–Gerlach experiment. The red handle is pointing to the right (thus choosing the θ_2-basis) and the ball takes, in both cases, the left path. Due to the magnet at the bifurcation, both balls will go the left path, resulting in a measurement spin left, $|\leftarrow\rangle$. If one chooses the right path, which stands for a Stern–Gerlach experiment in the θ_1-orientation, the magnet on the right-hand side guarantees a measurement outcome of spin "up", $|\nearrow\rangle$, in the θ_1-direction, as is demanded in Equations (9) and (10). Without communication at the bifurcation points, Alice and Bob will both measure spin "up", $|\nearrow\rangle$, when they chose the θ_1-orientated Stern–Gerlach experiment, which is forbidden in quantum mechanics.

Thus, the case distinction can be set up and experimentally performed by learners themselves. Afterward, the real quantum mechanical experiments and their results have to be shown and discussed so that the intuitive notion does not take root. Nevertheless, we consider this procedure to be useful because it shows clearly enough that the intuitive ideas reach their limits in quantum mechanics, but that the theory and the formalism provide us with tools that can describe quantum mechanical experiments exactly.

4. Conclusions

Quantum physics is generally perceived as one of the most difficult fields of physics, not because of an especially complex mathematical formalism but because of the rules that inherently govern quantum phenomena—rules that go directly against the intuition of how natural world works. Certainly, it is known that meaning can be brought to this formalism even for novices. However, this meaning still has to be explained to learners. For that reason, educators who have since been trying to help their students wrap their heads around these concepts often resorted to a multitude of representations, analogies, and thought experiments. Despite real experimental data of empirical evidence, quantum theory often seems out of reach—even more so in physics classrooms. This is why we have presented an approach to add to the multitude of representations by providing analog experiments as a platform for quantum physics education—yet another perspective to make experimental results more graspable.

Analog polarization modules bear the potential to adapt to various teaching scenarios and have already proven helpful in reconstructing fundamental principles of thinking about quantum objects. While expanding the methods and ways in which the models can be used, prototypes were already tested under real-world conditions: we collected feedback from students who showed keen interest in working with the physical simulations mimicking quantum physics experiments and started to have fun learning about the subject.

On the same level, in a different context, the analog Hardy experiment [25] achieves similarly promising results: disproving the existence of a local hidden parameter by simple means of comparison. The contradiction of quantum theoretical formalism and forbidden measurement is inherent to this classical (thus, local) experiment and is, therefore, predestined to be used in analogy-driven teaching. This model, too, has already received great acceptance in a physics classroom, albeit within a higher education setting in a course for advanced learners.

The next step is to build upon the results gained from testing the analog models, expand the corresponding teaching concept with further modules, and implement the findings into physics teaching by offering the materials we developed open source.

Author Contributions: All authors have contributed to conceptualization, methodology, validation, formal analysis, resources, and data curation; software, S.A.; investigation, S.A. and P.S.; writing—original draft preparation, S.A. and P.S.; writing—review and editing, H.C.; visualization, S.A. and P.S.; supervision, H.C. All authors have read and agreed to the published version of the manuscript.

Funding: This research received no external funding.

Data Availability Statement: Not applicable.

Conflicts of Interest: The authors declare no conflict of interest.

References

1. Einstein, A.; Podolsky, B.; Rosen, N. Can quantum-mechanical description of physical reality be considered complete? *Phys. Rev.* **1935**, *47*, 777–780. [CrossRef]
2. Müller, R. *Quantenphysik in der Schule*; Logos Verlag: Berlin, Germany, 2000.
3. Norsen, T. *Foundations of Quantum Mechanics: An Exploration of the Physical Meaning of Quantum Theory*; Springer International Publishing AG: Cham, Switzerland, 2017. Available online: https://link.springer.com/book/10.1007/978-3-319-65867-4 (accessed on 10 September 2022).
4. Bricmont, J. *Making Sense of Quantum Mechanics*; Springer International Publishing Switzerland: Cham, Switzerland, 2016. [CrossRef]
5. Dürr, D.; Lazarovici, D. *Understanding Quantum Mechanics: The World According to Modern Quantum Foundations*; Springer Nature Switzerland: Cham, Switzerland, 2020. [CrossRef]
6. Feynman, R.P.; Leighton, R.B.; Sands, M. *The Feynman Lectures on Physics, Volume III: Quantum Mechanics*; Basic Books: New York, NY, USA, 2011. Available online: https://feynmanlectures.caltech.edu/III_toc.html (accessed on 10 September 2022).
7. Greinert, F.; Müller, R.; Bitzenbauer, P.; Ubben, M.; Weber, K.-A. Requirements for future quantum workforce—A Delphi study. *J. Phys. Conf. Ser.* **2022**, *2297*, 012017.
8. Pospiech, G. *Quantencomputer & Co.: Grundideen und Zentrale Begriffe der Quanteninformation Verständlich Erklärt*; Springer Spektrum: Wiesbaden, Germany, 2021. [CrossRef]
9. Heusler, S.; Schlummer, P.; Ubben, M. The topological origin of quantum randomness. *Symmetry* **2021**, *13*, 581. [CrossRef]
10. Aehle, S.; Cartarius, H. Didaktische Ansätze für Quantum Random Number Generators (QRNG). *PhyDid B—Didakt. Phys.—Beiträge DPG-Frühjahrstagung* **2021**, *virtuell*, 489–493. Available online: https://ojs.dpg-physik.de/index.php/phydid-b/article/view/1146 (accessed on 10 September 2022).
11. Bitzenbauer, P.; Meyn, J. A new teaching concept on quantum physics in secondary schools. *Phys. Educ.* **2020**, *55*, 055031. [CrossRef]
12. Scheiger, P.; Nawrodt, R.; Cartarius, H. Interaktive und aktivierende Lehrkonzepte in der Theoretischen Physik. *PhyDid B-Didakt. Phys.—Beiträge DPG-Frühjahrstagung 2020, Bonn*, 77–83. Available online: https://ojs.dpg-physik.de/index.php/phydid-b/article/view/1166 (accessed on 10 September 2022).
13. Filk, T. *Quantenmechanik (Nicht Nur) für Lehramtsstudierende*; Springer Spektrum: Berlin/Heidelberg, Germany, 2019. [CrossRef]
14. Pospiech, G.; Schöne, M. Teacher education in quantum physics—A proposal for improving content knowledge. In *Electronic Proceedings of the ESERA 2019 Conference—The Beauty and Pleasure of Understanding: Engaging with Contemporary Challenges Through Science Education; Levrini, O., Tasquier, G.; Eds.; Bologna, Italy, 26–30 August 2019; Part 13: Pre-service Science Teacher Education*; Evagorou, M., Jimenez Liso, M.R., Eds.; Alma Mater Studiorium/University of Bologna: Bologna, Italy, 2020; pp. 1452–1461. Available online: https://www.esera.org/publications/esera-conference-proceedings/esera-2019 (accessed on 10 September 2022).

15. Weber, K.-A.; Friege, G.; Scholz, R. Quantenphysik in der Schule—Was benötigen Lehrkräfte? Ergebnisse einer Delphi-Studie. *Zeit. Didaktik Naturwis.—ZfDN* **2020**, *26*, 173–190. [CrossRef]
16. Ivanjek, L.; Shaffer, P.; Planinic, M.; McDermott, L. Probing student understanding of spectra through the use of a typical experiment used in teaching introductory modern physics. *Phys. Rev. Phys. Educ. Res.* **2020**, *16*, 010102. [CrossRef]
17. Greinert, F.; Müller, R. Playing with a quantum computer. *arXiv* **2021**, arXiv:2108.06271. [CrossRef]
18. Schlummer, P.; Laustöer, J.; Schulz-Schaeffer, R.; Abazi, A.; Schuck, C.; Pernice, W.; Heusler, S.; Laumann, D. MiReQu—Mixed Reality Lernumgebungen zur Förderung fachlicher Kompetenzentwicklung in den Quantentechnologien. *PhyDid B-Didakt. Phys.—Beiträge DPG-Frühjahrstagung* **2020**, *Bonn*, 451–459. Available online: https://ojs.dpg-physik.de/index.php/phydid-b/article/view/1044 (accessed on 10 September 2022).
19. Donhauser, A.; Bitzenbauer, P.; Meyn, J.-P. Von Schnee-und Elektronenlawinen: Entwicklung eines Erklärvideos zu Einzelphotonendetektoren. *PhyDid B—Didakt. Phys.—Beiträge DPG-Frühjahrstagung* **2020**, *Bonn*, 235–240. Available online: https://ojs.dpg-physik.de/index.php/phydid-b/article/view/1021 (accessed on 10 September 2022).
20. Novak, M.; Schwan, S. Does Touching Real Objects Affect Learning? *Educ. Psychol. Rev.* **2021**, *33*, 637–665. [CrossRef]
21. Chatterjee, H.J.; Hannan, L.; Thomson, L. An introduction to object-based learning and multisensory engagement. In *Engaging the Senses: Object-Based Learning in Higher Education*; Chatterjee, H.J., Hannan, L., Eds.; Routledge: New York, NY, USA, 2015; pp. 1–18. [CrossRef]
22. Moreno, R.; Mayer, R. Interactive Multimodal Learning Environments. *Educ. Psychol. Rev.* **2007**, *19*, 309–326. [CrossRef]
23. Kircher, E.; Girwidz, R.; Fischer, H. *Physikdidaktik: Grundlagen*; Springer Spektrum: Berlin/Heidelberg, Germany, 2020. [CrossRef]
24. Küblbeck, J.; Müller, R. *Die Wesenszüge der Quantenphysik: Modelle, Bilder, Experimente*; Aulis Verlag Duebne GmbH & Co KG: Cologne, Germany, 2003.
25. Hardy, L. Nonlocality for two particles without inequalities for almost all entangled states. *Phys. Rev. Lett.* **1993**, *71*, 1665–1668. [CrossRef] [PubMed]
26. Stadermann, H.K.E.; van den Berg, E.; Goedhart, M.J. Analysis of secondary school quantum physics curricula of 15 different countries: Different perspectives on a challenging topic. *Phys. Rev. Phys. Educ. Res.* **2019**, *15*, 010130. [CrossRef]
27. Müller, R.; Mishina, O. Milq—Quantum physics in secondary school. In *Teaching-Learning Contemporary Physics. Challenges in Physics Education*; Jarosievitz, B., Sükösd, C., Eds.; Springer: Cham, Switzerland, 2021; pp. 33–45. [CrossRef]
28. Passon, O.; Grebe-Ellis, J. Was ist eigentlich ein Photon? *Prax. Naturwis. (PdN)—Phys. Schule* **2015**, *64*, 46–48. Available online: https://www.physikdidaktik.uni-wuppertal.de/fileadmin/physik/didaktik/Forschung/Publikationen/Passon/Passon_Grebe-Ellis_2015_Moment_mal_was_ist_eigentlich_ein_Photon.pdf (accessed on 10 September 2022).
29. Merzbacher, E. *Quantum Mechanics*; Wiley International Edition: New York, NY, USA, 1970.
30. Tim Maudlin, J. What Bell did. *Phys. A Math. Theor.* **2014**, *47*, 424010. [CrossRef]
31. Goldstein, S.; Norsen, T.; Tausk, D.V.; Zanghi, N. Bell's theorem. *Scholarpedia* **2011**, *6*, 8378. [CrossRef]
32. Bell, J.S. On the Einstein—Podolsky—Rosen paradox. *Phys. Phys. Fiz.* **1964**, *1*, 195–290. [CrossRef]

physics

MDPI

Opinion

Classical Limits of Light Quanta

Clara Valeria Fuchs [1] and Thomas Filk [2,*]

[1] Hans-Thoma-Gymnasium Lörrach, 79540 Lörrach, Germany
[2] Institute of Physics, University of Freiburg, Hermann-Herder-Str. 3, 79104 Freiburg, Germany
* Correspondence: thomas.filk@physik.uni-freiburg.de

Abstract: It is argued that from a formal point of view, the classical limit of light quanta or photons is not that of a point-like particle but that of a geometric ray. According to this view, standard particle-wave dualism, which is often used in schools to describe the quantum behavior of massive objects, could be replaced by a ray-wave dualism (or even a particle-ray-wave trialism), which seems to be more appropriate for massless quantum objects such as photons. We compare the limits leading from quantum electrodynamics to a classical (Hamiltonian) theory of particles for electrons with those leading from photons via Maxwell's equations to geometric ray optics. We also discuss the question to which extent Maxwell's theory for electromagnetic waves should be considered as being on the same formal level as Schrödinger's or Dirac's theory.

Keywords: quantum theory; classical limits; photon; didactics of physics

1. Introduction

There exist many different didactical approaches to introducing quantum theory in school. A common one is emphasizing the wave-like nature of electrons, e.g., in double slit experiments. Another one is pointing out the particle-like nature of light, e.g., when light of very low intensity is detected on scintillator screens or with high-resolution charged-couple device (CCD) cameras, or in the photoelectric effect or in Compton scattering. The wave-particle duality of any form of matter is introduced as one of the characteristic features of quantum theory.

However, in their educational curriculum, pupils are commonly introduced to ray optics first, before they are confronted with wave optics. Rays describe light in a form that appeals to our daily experience. Therefore, one of the naturally arising questions is, whether or not rays are a more appropriate model for "classical photons" instead of emphasizing their particle-like nature. It has been noted that the notion of photons as particles can lead to gross misconceptions; see [1–5].

This raises another important issue: What is the classical limit of photons—optical rays or Maxwell's theory of electromagnetism, i.e., essentially wave optics? While Maxwell's theory is considered "classical physics" for many reasons (some of them are summarized in Section 5), ray optics is a particular limit of this classical theory. However, on a formal mathematical level, Maxwell's theory should be compared to Schrödinger's or Dirac's theory of electrons, which are unequivocally quantum theories of electrons; see Figure 1.

Figure 1 essentially represents the content of this paper in a schematic form. On the highest level, we have quantum electrodynamics (QED), a quantum field theory. This is the level of second quantization, in which the Hilbert space is often represented as a Fock space (i.e., a space where the base states have definite particle numbers), and it is on this level that one can speak of electrons and photons as quantum objects. The dynamical variables are field operators (to be precise, these operators have to be smeared out with test functions, i.e., they are operator-valued distributions).

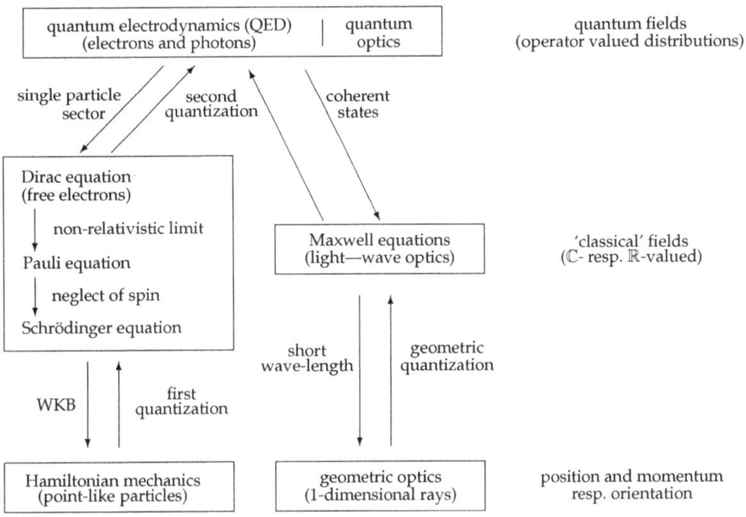

Figure 1. Comparison of the various levels of models for electrons and photons and the approximations from higher to lower levels, as well as the "quantization" procedures from lower to higher levels. For details, see text.

We place quantum optics on the same level, even though in standard applications QED and quantum optics look very different: While QED is more concerned with the scattering amplitudes of electrons, uses Dirac's interaction picture and perturbation theory as its main analytical tools, and treats photons mostly as virtual exchange particles, quantum optics prefers the Fock space representation, a non-Lorentz invariant choice of gauge, and deals mainly with non-classical properties of photon states, replacing electrons often by effective two-state systems which can absorb or emit photons.

On the middle level we have field theories. The dynamical degrees of freedom are real or complex-valued fields: the gauge field of electrodynamics and the spinor fields of Dirac's theory of electrons. While Dirac's theory for the electron (and its anti-particle, the positron) is unequivocally considered a quantum theory, Maxwell's theory for the electromagnetic field is generally considered a classical (field) theory. From a mathematical viewpoint, both theories are on the same level. However, from the viewpoint of physics and the physical interpretation of the fields, there are many differences, which are considered in Section 5. On the same formal level as Dirac's theory for the electron is Schrödinger's equation, which can be derived from Dirac's equation by taking a non-relativistic limit and neglecting the spin degree of freedom (if spin is included, one arrives at the Pauli equation).

The route from the middle level (field theory) to the highest level (quantum field theory) is second quantization: fields are replaced by operators satisfying canonical commutation relations. The way back from quantum field theory to field theory (i.e., from field operators to real or complex-valued fields), involves certain approximations: for electrons the restriction to the one-particle sector of the Fock space and for photons the restriction to so-called coherent states, i.e., eigenstates of the (annihilation part of) field operators. These relationships are discussed in more detail in the next section (Section 2).

On the lowest level, we find for the electrons a Newtonian (or rather Hamiltonian) theory of point-like particles. The classical degrees of freedom are position, $\boldsymbol{x}(t)$, and momentum, $\boldsymbol{p}(t)$, variables, which make up the classical phase space of particles. The step from here to Schrödinger's equation is (first) quantization: replacing the position and momentum variables by operators $\hat{Q}$ and $\hat{P}$ satisfying canonical commutation relations, replacing the classical Hamilton function $H(x,p)$ by a Hamilton operator $\hat{H}(\hat{Q},\hat{P})$ and postulating Schrödinger's equation for wave functions $\psi(x)$, which represent quantum states (throughout this paper operators are marked with a hat ˆ). The other direction—the route from Schrödinger's equation to Hamiltonian mechanics, involves a certain approximation (the Wentzel–Kramers–Brillouin (WKB) approximation), which is essentially a short-wave approximation for the Schrödinger fields.

In a similar way, one can perform a short-wave approximation of the wave equation (or Maxwell's theory) for electromagnetic fields and arrive at geometrical optics or ray optics. This is the main reason why we propose to consider rays as the classical limit of light quanta and not point-like particles. There is no natural limit from Maxwell's theory to point-like particles. The transitions between the middle-level and lowest level, i.e., from Schrödinger to Newton on the one hand and from wave optics to ray optics, on the other hand, are considered in Section 3. What is less known is that geometric optics has a symplectic structure, similar to the phase space of Hamiltonian mechanics, and can be quantized. The quantization of ray optics leads to wave optics, i.e., back to the middle level. This is sketched in Section 4.2.

Section 5 deals with the question of whether Maxwell's theory is a quantum theory or a classical theory. Of course, eventually, this is a question of definitions and there are many convincing physical arguments to label Maxwell's theory a classical theory, however, there are also justified formal arguments to consider it as the quantized theory of geometric optics.

Finally, the conclusions (Section 6) summarize the arguments why the classical limit of photons may not only be described by a "particle-wave" duality, i.e., photons as particles (this model is often used to explain the photoelectric effect or Compton scattering) or as electromagnetic waves (explaining, e.g., interference and diffraction), but that there is a third option: classical photons as geometric rays, for example, when one considers light rays in optical systems such as lenses or mirrors.

Sections 2–4.2 are "technical sketches": they indicate the technical steps and relations between the various levels and theories. For more detailed information, the reader is referred to the literature. On the other hand, readers who are happy with the introductory remarks indicating the various relationships may also skip these sections and directly proceed to the last two sections.

2. Electrons, Photons, and the Electromagnetic Field

The starting point of our considerations is the Lagrange density of electrodynamics:

$$\mathcal{L} = -\frac{1}{4}F_{\mu\nu}F^{\mu\nu} + \bar{\psi}(x)\left[\gamma^\mu(\mathrm{i}\partial_\mu + eA_\mu(x)) - m\right]\psi(x)\,. \tag{1}$$

Here, $F_{\mu\nu} = \partial_\mu A_\nu - \partial_\nu A_\mu$ is the field tensor with the electric and magnetic fields $\mathbf{E}$ and $\mathbf{B}$ as components. The indices μ, $\nu = 0, 1, 2$, and 3, denote the time (0) and space coordinates. A_μ is the gauge field, $\psi(x)$ is a 4-component spinor field and $\bar{\psi}(x) = \psi^\dagger\gamma^0$ the adjoint spinor (with $\psi^\dagger$ the complex conjugated row vector corresponding to ψ), γ^μ are the γ-matrices satisfying anti-commutation relations $\{\gamma^\mu, \gamma^\nu\} = \eta^{\mu\nu}$ where $\eta^{\mu\nu}$ is the Minkowski bilinear form. e is the electric charge, m is the particle mass, and $\partial_\mu = \partial/\partial x_\mu$. In this Section only, the units are such that the Planck's reduced constant $\hbar = 1$ and the speed of light $c = 1$. The corresponding equations of motion (using Lorentz gauge $\partial_\mu A^\mu = 0$) are:

$$\Box A_\mu(x) = e\bar{\psi}(x)\gamma_\mu\psi(x) \qquad \text{and} \qquad \gamma^\mu(\mathrm{i}\partial_\mu - eA_\mu)\psi(x) - m\psi(x) = 0\,. \tag{2}$$

The first equation is essentially Maxwell's equation expressed in terms of the gauge field and the electric current $j_\mu = e\bar{\psi}(x)\gamma_\mu\psi(x)$, the second equation is Dirac's equation.

2.1. Quantum Electrodynamics

The quantization of this field theory leads to QED: The fields are replaced by operators satisfying the above equations of motion and certain commutation relations. The free fields, in terms of which the theory is formulated perturbatively, can be expressed in terms of creation and annihilation operators in the following form:

$$\begin{aligned}
\hat{A}_\mu(x) &= \frac{1}{\sqrt{(2\pi)^3}} \int_{\mathbb{R}^3} \frac{d^3\boldsymbol{k}}{\sqrt{2\omega_k}} \sum_{\alpha=1}^{2} \epsilon_{\alpha\mu} \left(\hat{a}_\alpha(\boldsymbol{k}) e^{ikx} + \hat{a}_\alpha^\dagger(\boldsymbol{k}) e^{-ikx} \right) \\
\hat{\Psi}(x) &= \frac{1}{\sqrt{(2\pi)^3}} \int_{\mathbb{R}^3} \frac{d^3\boldsymbol{k}}{\sqrt{2\omega_k}} \sum_{s=\pm} \left(\hat{b}_s(\boldsymbol{k}) u(\boldsymbol{k},s) e^{ikx} + \hat{d}_s^\dagger(\boldsymbol{k},s) v(\boldsymbol{k},s) e^{-ikx} \right) \\
\hat{\Psi}^\dagger(x) &= \frac{1}{\sqrt{(2\pi)^3}} \int_{\mathbb{R}^3} \frac{d^3\boldsymbol{k}}{\sqrt{2\omega_k}} \sum_{s=\pm} \left(\hat{b}_s^\dagger(\boldsymbol{k}) \bar{u}(\boldsymbol{k},s) e^{-ikx} + \hat{d}_s(\boldsymbol{k},s) \bar{v}(\boldsymbol{k},s) e^{ikx} \right) \\
& k = (\omega_k, \boldsymbol{k}), \quad \omega_k = \sqrt{\boldsymbol{k}^2 + m^2}, \quad x = (t, \boldsymbol{x}) \\
& kx = \boldsymbol{k} \cdot \boldsymbol{x} - \omega_k t,
\end{aligned}$$

where $u(\boldsymbol{k},s), v(\boldsymbol{k},s)$ are solutions of Dirac's equation and $\epsilon_{\alpha\mu}$ are two polarization vectors (the technical details are slightly more complex: in order to keep Lorentz invariance, one has to choose four polarization vectors (i.e., $\alpha = 0, \ldots, 3$) and, correspondingly, four annihilation and creation operators. In the end, one has to show that two of these—corresponding to a longitudinal photon and a pure time-like photon—do not contribute. However, such technical details shall not concern us here and the interested reader is referred to the standard textbooks on quantum field theory) of the gauge fields (orthonormal to each other and orthogonal to the momentum $\boldsymbol{k}$). $\hat{a}_\alpha(\boldsymbol{k})$ and $\hat{a}_\alpha^\dagger(\boldsymbol{k})$ are annihilation and creation operators for field modes with momentum $\boldsymbol{k}$ and polarisation ϵ_α, respectively. These field modes created by $\hat{a}_\alpha(\boldsymbol{k})^\dagger$ are called photons. In a similar way, $\hat{d}_s^\dagger(\boldsymbol{k})$ and $\hat{b}_s^\dagger(\boldsymbol{k})$ are creation operators for positrons and electrons, respectively, $\hat{d}_s(\boldsymbol{k})$ and $\hat{b}_s(\boldsymbol{k})$ are the corresponding annihilation operators. The index s accounts for the spin degree of freedom. m is zero for the gauge field and equals the electron mass for the spinor fields.

If we denote by $|\Omega\rangle$ the vacuum state of the theory, i.e., the state of lowest energy, then

$$|\alpha^\mu\rangle = \int_{\mathbb{R}^4} d^4x\, \alpha^\mu(x) \hat{A}_\mu^+(x) |\Omega\rangle \tag{3}$$

$$\text{with} \quad \hat{A}_\mu^+(x) = \frac{1}{\sqrt{(2\pi)^3}} \int_{\mathbb{R}^3} \frac{d^3\boldsymbol{k}}{\sqrt{2\omega_k}} \sum_{\alpha=1}^{2} \epsilon_{\mu\alpha} \left(\hat{a}_\alpha^\dagger(\boldsymbol{k}) e^{-ikx} \right) \tag{4}$$

describes a one-photon state with wave function $\alpha^\mu(x)$. In a similar way

$$|\psi\rangle = \int_{\mathbb{R}^4} d^4x\, \psi(x) \hat{\Psi}^+(x) |\Omega\rangle \tag{5}$$

$$\text{with} \quad \hat{\Psi}^+(x) = \frac{1}{\sqrt{(2\pi)^3}} \int_{\mathbb{R}^3} \frac{d^3\boldsymbol{k}}{\sqrt{2\omega_k}} \sum_{s=\pm} \left(\hat{b}_s^\dagger(\boldsymbol{k},s) u(\boldsymbol{k},s) e^{-ikx} \right) \tag{6}$$

describes a single electron state with wave function $\psi(x)$.

All these expressions are quite formal and many technical details are left out. The interested reader is referred to the standard literature on quantum field theory and second quantization (e.g., [6–10]).

2.2. Quantum Optics

We consider quantum optics as a special branch of QED. In particular, equations such as Equations (3) and (4) remain true. In contrast to QED, which in its standard applications concentrates on n-point functions and scattering amplitudes for electrons and positrons, quantum optics is mainly concerned with the non-classical behavior of photonic states, i.e., with deviations of photonic states from the predictions of Maxwell's theory. This non-classical behavior is mostly due to the bosonic statistics of photons—a paradigm being the Hong-Ou-Mandel effect [11]—or due to the discrete nature of photons (see, e.g., [12]). A related application consists of the non-classical properties of intensity correlation functions (see, e.g., [13]).

Closest to the classical electromagnetic field are so-called "coherent" states, which are described by a Poisson-distribution in photon number. For simplicity, we describe these states for a given mode and polarization only, i.e., for a single harmonic oscillator.

Let us define the displacement operator

$$\hat{D}(\alpha) = \exp\left(\alpha\hat{a}^\dagger - \alpha^*\hat{a}\right). \tag{7}$$

Acting on the vacuum state yields

$$|\alpha\rangle = \hat{D}(\alpha)|\Omega\rangle, \tag{8}$$

which is an eigenstate of the annihilation operator $\hat{a}$:

$$\hat{a}|\alpha\rangle = \alpha|\alpha\rangle, \tag{9}$$

and for the probability w_n of finding n photons in such a state, one obtains a Poisson distribution:

$$w_n = \exp\left(-|\alpha|^2\right)\frac{|\alpha|^{2n}}{n!}. \tag{10}$$

$|\alpha\rangle$ represents a state, for which the amplitude is given by a Gaussian distribution around the value α. It is like the ground state of a harmonic oscillator (which is a Gaussian distribution around 0) shifted to the value α. In a formal classical limit these states become sharply distributed around their classical values.

Coherent states often have similar properties as single photon states. In particular, the superposition principle is the same: The superposition of two single-photon states with, say, different polarizations or originating from different slits in a double slit experiment yields a single-photon state with the superimposed polarization or the superimposed wave-function. The same relations hold for coherent states. Only when the discreteness of photon states becomes relevant is the difference obvious (see, e.g., [12]).

Again, for details, the interested reader is referred to the literature (e.g., [14,15]).

3. The Classical Limits of Field Theory

The route from field theory to the classical requires several steps. The Dirac equation is a relativistic equation for electrons and their antiparticles, the positrons, while the Schrödinger equation is a non-relativistic equation for (spinless) electrons. The intermediate steps are (i) a non-relativistic approximation for the solutions of Dirac's equation, which leads to the Pauli equation, and (ii) the neglect of the spin degree of freedom leading to Schrödinger's equation. From Schrödinger's equation there are several (mostly equivalent) approximations, the most known being the WKB approximation, leading to Newtonian mechanics and the classical picture of an electron as a point particle.

The photon part of field theory is essentially Maxwell's theory. Neglecting the "spin" (i.e., the polarization) leads to the wave equation for the gauge field and, essentially, wave optics. A non-relativistic limit is not possible here, mainly because the photon is a massless particle which propagates with the velocity of light. Quite often the opposite limit, $c \to \infty$ is taken and any temporal dependence eliminated. However, the "classical" limit here corresponds to a short-wavelength approximation, leading to geometrical optics.

These limits are briefly described in the following sections. For technical details, the reader is again referred to the literature: for the steps from Dirac to Schrödinger, e.g., [6], for the WKB approximation, e.g., [16–18], and for the short-wave approximation, e.g., [15].

3.1. From Dirac to Schrödinger

If we multiply the Dirac equation with γ_0 it assumes formally the form of a Schrödinger equation

$$\mathrm{i}\hbar\frac{\partial}{\partial t}\psi(x) = \hat{H}\psi(x) \qquad \text{with} \qquad \hat{H} = \boldsymbol{\gamma}\cdot(c\,\mathrm{i}\hbar\nabla + e\boldsymbol{A}) + \gamma_0 mc^2 + e\phi\,. \tag{11}$$

Here, $A_\mu \simeq (\phi, \boldsymbol{A})$ and $\gamma_\mu \simeq (\gamma_0, \boldsymbol{\gamma})$. In a base where γ_0 is diagonal, the solutions for small momenta of the four components of the Dirac spinor $\psi(x)$ can be associated with the electron (two components taking care of the spin and the spatial degrees of freedom) and the positron, respectively. By making the ansatz $\psi(x) \to \tilde{\psi}(x) = \mathrm{e}^{-\mathrm{i}mc^2t/\hbar}\psi(x)$, which essentially shifts the energy scale for the new spinor $\tilde{\psi}(x)$ to the rest energy $E_\mathrm{e} = m_\mathrm{e}c^2$ of the electron, and neglecting contributions to the energy, which are of the order $2m_\mathrm{e}c^2$, one obtains the Pauli equation for two of the components:

$$\mathrm{i}\hbar\frac{\partial}{\partial t}\tilde{\psi}(x) = \left(\frac{1}{2m}\left(\mathrm{i}\hbar\nabla - \frac{e}{c}\boldsymbol{A}(x)\right)^2 + e\phi(x) + \mu_\mathrm{B}\boldsymbol{\sigma}\cdot\boldsymbol{B}(x)\right)\tilde{\psi}(x)\,. \tag{12}$$

σ_i are the Pauli matrices. The other two components of $\tilde{\psi}$ effectively vanish (in the sense of distributions, i.e., they oscillate so rapidly due to their high energies that they do not give contributions when smeared with 'low energy' test functions). If the external magnetic field vanishes, we obtain Schrödinger's equation:

$$\mathrm{i}\hbar\frac{\partial}{\partial t}\tilde{\psi}(x) = \left(-\frac{\hbar^2}{2m}\Delta + e\phi(x)\right)\tilde{\psi}(x)\,. \tag{13}$$

Now $\tilde{\psi}$ can be considered a one-component complex field.

3.2. From Schrödinger to Newton

There are several, largely equivalent routes from Schrödinger's equations to Newton's equations or rather to Hamiltonian mechanics. One method is the WKB approximation. One way to think of WKB is that the wavelength λ associated to a given momentum $\boldsymbol{p}$ by deBroglie's relation $|\boldsymbol{p}| = h/\lambda$ is small compared to distances, over which the potential $V(x)$ varies, so that locally one has a box-potential. So, effectively, this is a small wave-length approximation. Making the following ansatz for the wave function,

$$\psi(x) = \exp\left(\frac{\mathrm{i}}{\hbar}S(\boldsymbol{x},t)\right), \tag{14}$$

and inserting this ansatz into Schrödinger's equation, one obtains the following differential equation for $S(\boldsymbol{x},t)$:

$$-\frac{\partial S}{\partial t} = \frac{1}{2m}\left((\nabla S)^2 - \mathrm{i}\hbar\Delta S(\boldsymbol{x},t)\right) + V(\boldsymbol{x})\,. \tag{15}$$

If S is a slowly varying function (or, less formally, if we neglect terms proportional to $\hbar$), this becomes the classical Hamilton–Jacobi equation:

$$\frac{\partial S_0}{\partial t} + \frac{1}{2m}(\nabla S_0)^2 + V(\boldsymbol{x}) = 0\,. \tag{16}$$

The classical trajectories are orthogonal to the planes of constant phase and are determined by

$$\boldsymbol{p}(t) = \nabla S(\boldsymbol{x}(t),t) \qquad \text{or} \qquad \frac{\mathrm{d}\boldsymbol{x}(t)}{\mathrm{d}t} = \frac{1}{m}\nabla S(\boldsymbol{x}(t),t)\,. \tag{17}$$

For the position $\boldsymbol{x}(t)$, one obtains (see, e.g., [19]):

$$m\frac{\mathrm{d}^2\boldsymbol{x}}{\mathrm{d}t^2} = -\boldsymbol{\nabla}V(\boldsymbol{x}(t))\,. \tag{18}$$

3.3. From Maxwell to Ray Optics

In complete analogy to the WKB approximation for Schrödinger's equation, one can make the ansatz,

$$\boldsymbol{E}(\boldsymbol{x},t) = \boldsymbol{e}(\boldsymbol{x})\mathrm{e}^{\mathrm{i}k(S(\boldsymbol{x})-ct)}\,, \qquad k = \frac{2\pi}{\lambda}\,, \tag{19}$$

for the electric field (and a similar ansatz with the same phase for the magnetic field). For large values of k (or small wave-lengths λ), one can make a formal expansion of the phase $S(\boldsymbol{x})$ and the amplitude $\boldsymbol{e}(\boldsymbol{x})$ in powers of $1/k$. For the leading term S_0 in this expansion, Maxwell's equations lead to the equation

$$\boldsymbol{\nabla}S_0(\boldsymbol{x}))^2 = n^2(k,\boldsymbol{x})\,, \tag{20}$$

where $n(k,\boldsymbol{x})$ is the wavelength-dependent optical density of the medium. This is the famous eikonal equation leading to geometrical optics and Fermat's principle. One obtains trajectories of rays as the solutions of

$$\frac{\mathrm{d}\boldsymbol{x}(s)}{\mathrm{d}s} = \frac{1}{n(k,\boldsymbol{x}(s))}\boldsymbol{\nabla}S_0(\boldsymbol{x}(s))\,, \tag{21}$$

where the arc-length parametrization is used. This equation should be compared with Equation (17). For more details see, e.g., [15].

4. The Quantization of Hamiltonian Mechanics and Ray Optics

"Quantization" of Newtonian mechanics leads to quantum mechanics and Schrödinger's equation for the quantum states of, e.g., electrons. This is the subject of any course on quantum mechanics. However, ray optics can be quantized along similar lines leading to wave optics. This is less known and we briefly sketch the idea, referring the interested reader to the literature (e.g., [20]).

4.1. A Different View Onto Quantization of Hamiltonian Mechanics

The state space of Hamiltonian classical mechanics, i.e., the phase space, has a natural symplectic structure. This structure reveals itself, e.g., in the Poisson brackets, with the fundamental relation $\{q,p\} = 1$. One way to formulate the quantization procedure is to replace q and p (generalized position and momentum) by operators $\hat{Q}$ and $\hat{P}$ and to require these operators to satisfy canonical commutation relations $[\hat{Q},\hat{P}] = \mathrm{i}\hbar$.

There is, however, a different perspective onto the same procedure. The Hamiltonian equations of motion define symplectic diffeomorphisms ϕ_t, i.e., area preserving differentiable mappings of phase space onto itself: to each point (q,p) of phase space is associated the point $\phi_t(q,p) = (q(t),p(t))$, i.e., the point in phase space, where an object, which initially was at (q,p), will be after a time t. Liouville's theorem tells us that volumes in phase space will be preserved. These symplectic diffeomorphisms have a monoid structure with respect to t in the sense that $\phi_{t_2}\circ\phi_{t_1} = \phi_{t_2+t_1}$ and $\phi_0 = \mathbf{1}$. These diffeomorphisms are generated by Hamilton's equations of motion: $(\dot{q},\dot{p}) = (\partial_p H, -\partial_q H)$.

Quantization can now be formalized by looking for unitary representations of this symplectic diffeomorphisms on a Hilbert space. In quantum mechanics these are the unitary time evolution operators $\hat{U}(t) = \exp\left(-\frac{\mathrm{i}}{\hbar}\hat{H}t\right)$. They satisfy Schrödinger's equation

$$\mathrm{i}\hbar\frac{\mathrm{d}}{\mathrm{d}t}\hat{U}(t) = \hat{H}\hat{U}(t) \qquad \hat{U}(0) = \mathbf{1}\,. \tag{22}$$

4.2. Wave Optics as the Quantization of Ray Optics

Let us first show that geometric optics can also be given a symplectic structure. Then, we indicate how this symplectic structure can be quantized.

Let us denote by $x(z)$ the point, where a geometric ray punctures a plane orthogonal to the z-direction, which is taken to be the propagation direction of the ray. Furthermore, we denote by $p(z)$ the tangent of the angle (between the z- and the x-direction) of the ray (see Figure 2). We can now represent the configuration of a ray by a two-dimensional vector $\begin{pmatrix} x(z) \\ p(z) \end{pmatrix}$. For simplicity, we consider this as a one-dimensional problem. In general, $x(z)$ and $p(z)$ will each be two-dimensional vectors (the location in the plane at point z in the propagation direction of the ray and the two tangents specifying the orientation of the ray at this plane) leading to a four-dimensional formalism.

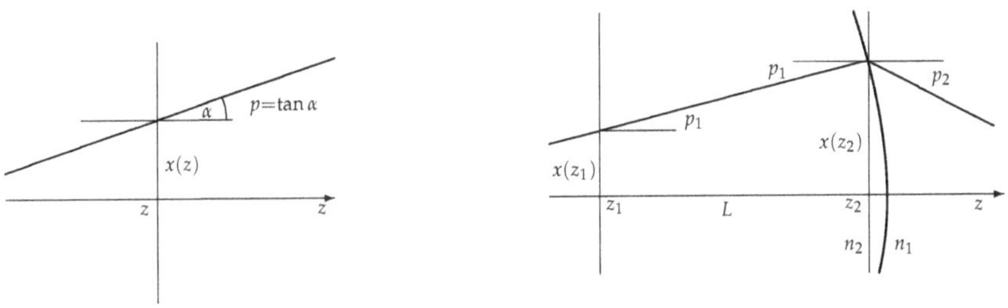

Figure 2. (**Left**): Characterization of a light ray by the parameters x and p. This refers to one dimension. In a plane orthogonal to the z-direction one has two parameters for the location and two parameters for the orientation of the ray at a position z. (**Right**): The transformations "translation" (T) by a length L and "defraction" (D) at an interface between two different optical densities n_1 and n_2. For details see text.

Essentially, there are two transformations which a ray can be subject to: (T) a "translation" describes the change in parameters when the ray simply propagates for a certain distance L without being deflected, and (D) a "defraction" of the ray, e.g., at an interface between two different optical densities n_1 and n_2. Both transformations can be described by a matrix:

$$T = \begin{pmatrix} 1 & L \\ 0 & 1 \end{pmatrix}, \qquad D = \begin{pmatrix} 1 & 0 \\ B & 1 \end{pmatrix}, \tag{23}$$

where L is the distance by which the ray propagates in a medium of constant optical density, and $B = (n_2 - n_1)/R$ is the "refractivity" of the transition from an optical medium with density n_1 to an optical medium with optical density n_2, while R is the radius of the curvature of the interface between these two media.

Both matrices, T and D, are symplectic: they have determinant 1 and a similarity transformation with these matrices leaves the antisymmetric matrix $J = \begin{pmatrix} 0 & 1 \\ -1 & 0 \end{pmatrix}$ invariant. Therefore, one finds a symplectic structure in geometric optics.

There is a general procedure to quantize spaces with a symplectic structure. Again, one looks for infinite dimensional unitary representations of the symplectic propagation matrices in a Hilbert space of square integrable functions. The time parameter t is now replaced by the propagation axis z (formally this corresponds to a limit $c \to \infty$ such that a time parameter is not involved; the transition to a relativistic formulation with finite c is an additional structure—in parallel to the transition from the non-relativistic Schrödinger equation to a relativistic Dirac formalism). The two matrices T and D are represented by unitary operators $\hat{\rho}(T)$ and $\hat{\rho}(D)$ for which the action on a square integrable function

$f(x;z)$ (square integrable in the argument x; z is an external parameter like t in quantum mechanics) is defined by

$$(\hat{\rho}(T)f)(x;z+L) = \exp\left(-\frac{i\pi n}{4}\right)\sqrt{\frac{k}{2\pi L}}\int \exp\left(ik\frac{(x-y)^2}{2L}\right)f(y;z)\,dy \quad (24)$$

$$(\hat{\rho}(D)f)(x;z) = \exp\left(-ik\frac{xBx}{2}\right)f(x;z)\,, \quad (25)$$

k being an open parameter. Note that $\hat{\rho}(T)$ essentially represents the unitary time evolution operator for a free particle in quantum mechanics with t replaced by z (and k by $\hbar$). In quantum mechanics, $\hat{\rho}(D)$ would correspond to a collision where the momentum is changed abruptly. One can easily show that these two unitary operators satisfy the same relations as the matrices T and D and, therefore, are a representation of these symplectic transformations. The applications of these formulae to various apertures leads to the integrals of Fresnels's wave optics (for details, see, e.g., [20]).

Note that the wavelength appears as a free parameter in this quantization procedure, similar to Planck's constant in quantum mechanics. Geometric ray optics does not contain the wavelength of light. Vice versa, the wavelength disappears in the leading terms of a short-wavelength approximation, which essentially is an expansion in terms of the wavelength, in a similar way as $\hbar$ disappears in a WKB approximation, which can be formulated as an expansion in terms of $\hbar$.

5. Maxwell's Theory—A Classical Theory of Electromagnetic Fields or a Quantum Theory of Ray Optics?

There are many arguments in favor of Maxwell's theory being a classical theory. Of course, as long as "classical theory" is not defined, the header question remains meaningless. Furthermore, arguably an "and" would be more appropriate in the header of this section than an "or". We list a few arguments pro and contra the assignment of "classical" or "quantum" to Maxwell's theory.

5.1. Maxwell's Equations Do Not Contain Planck's Constant

This is presumably the argument most often used by protagonists of Maxwell's theory being a classical theory. In a quantum theory $\hbar$ should appear.

The main reason that $\hbar$ does not appear in Maxwell's theory is that we are dealing with a relativistic theory of massless objects. Energy E and momentum $\boldsymbol{p}$ appear with the same power in the relativistic relation,

$$E^2 = m^2c^4 + \boldsymbol{p}^2c^2\,. \quad (26)$$

Replacing E and $\boldsymbol{p}$ by ω and $\boldsymbol{k}$ using deBroglie's relations yields:

$$\hbar^2\omega^2 = m^2c^4 + \hbar^2\boldsymbol{k}^2c^2\,, \quad (27)$$

and dividing this equation by $\hbar^2$ leaves Planck's constant only in the mass term (which essentially becomes Compton's wavelength for an object of mass m). For $m = 0$, Planck's constant vanishes from the equations. In other words, the (massless) wave equation,

$$\frac{1}{c^2}\frac{\partial^2 A_\mu(x)}{\partial t^2} - \Delta A_\mu(x) = 0\,, \quad (28)$$

contains no $\hbar$, because the $\hbar$-terms drop out of the dispersion relation

$$\frac{1}{c^2}\hbar^2\omega^2 = \hbar^2 k^2\,. \quad (29)$$

For the same reason, Dirac's equation for massless spin-$\frac{1}{2}$ fermions contains no $\hbar$. Does that make massless neutrinos (as in the standard model) classical objects?

In a non-relativistic theory such as Schrödinger's theory, the energy-momentum dependence is not homogeneous: it is the classical relation

$$E = \frac{1}{2m}\boldsymbol{p}^2 + V(\boldsymbol{x}), \tag{30}$$

and using deBroglie's relations necessarily leaves us with $\hbar$s:

$$\hbar\omega = \frac{\hbar^2}{2m}\boldsymbol{k}^2 + V(\boldsymbol{x}), \tag{31}$$

which becomes Schrödinger's equation when we require this relation to hold for a wave $\psi(\boldsymbol{x})$ of frequency ω and wave vector $\boldsymbol{k}$.

5.2. The Fields in Maxwell's Theory Have no Interpretation in Terms of a Born Rule

The interpretation of fields (neither the electromagnetic fields $\boldsymbol{E}$ and $\boldsymbol{B}$ nor the gauge field A_μ) is not that of "a probability amplitude of finding a photon". In other words, the absolute squares of these fields do not define a probability density for events in analogy to the Born rule in quantum mechanics. In fact, it can be shown (see, e.g., [21]) that no such relativistically invariant probability density can be constructed from these fields. There are several differences between the electronic part and the photonic part of the field theory which are responsible for this.

One of the reasons is that the relation between energy E and "number of photons" n depends on the frequency: $E = n\hbar\omega$. Furthermore, states which have a well-defined electric or magnetic field do not have a well-defined photon number. Indeed, the coherent states in quantum optics have a Poisson distribution of photon number. Expressed differently, the field operators and the number operator for photons do not commute.

On the other hand, there are relationships between the gauge field $\boldsymbol{A}$ or the electric field $\boldsymbol{E}$ and the number of photons (see, e.g., [16]). When we are dealing with fixed frequencies and consider only relative quantities (i.e., not absolute normalizations), we can interpret the absolute square of the gauge potential or the electric field as "relative frequency" and, eventually, as a probability. This holds in particular for polarization experiments, where the vector character of the electric field or the vector potential can be used as a two-dimensional probability amplitude and the absolute square (properly normalized) as a probability for "detecting an event".

5.3. Many Systems of Single Objects vs. One System of Many Objects

Quantum mechanics makes predictions about probabilities, at least in the Copenhagen interpretation. One can test such predictions by measuring relative frequencies. Ensemble interpretations of quantum mechanics claim that quantum states refer to ensembles—equally prepared systems—only [22]. The interference fringes of electrons in a double slit experiment [23] are obtained by letting many single electrons prepared with the same momentum pass through a double slit, preferably in such a way that they cannot influence each other, i.e., one at a time.

One might argue that for photons it should be possible that a single system with many photons prepared with the same quantum numbers might also allow to test probability predictions as relative frequencies. Photons are bosons and many photons can be in the same quantum state, which is not possible for electrons because of the Pauli principle. Furthermore, in contrast to electrons, photons do not interact (quantum corrections can effectively lead to photon-photon interactions via virtual electron-positron loops, but these effects are negligible and do not spoil the argument). Therefore, a single multi-particle state of photons with identical quantum numbers allows for the measurement of relative frequencies of photons in form of relative intensities. This might replace the preparation of ensembles of many single-particle states of photons. Even though correct in principle, the question remains to which extent states of classical fields in Maxwell's theory really represent such states of identical photons.

As mentioned above, this is not the case. States with more or less sharply defined field strengths correspond to coherent states of many photons. These states are superpositions of photon states with an arbitrary number of photons and do not describe a many-photon state with a definite photon number—they are not eigenstates of the photon number operator. Repeated measurements of relative frequencies in such systems do not yield the same numbers. This refers to the coherent states of laser light. Thermal light is even more diverse. Up to now, there are no good methods to prepare many-photon states with macroscopic numbers of photons in eigenstates of the number operator. Such states would exhibit an extreme quantum nature and they would differ from the states corresponding to well-defined field strengths in Maxwell's theory.

Essentially, this leads to the same conclusion as Section 5.2: the electromagnetic fields or the vector potentials do not represent wave functions for identically prepared photons, for which the integral over the absolute square has the interpretation of relative frequencies of finding definite numbers of photons in a well-defined volume. However, in the limit of large photon number N, where fluctuations can be neglected and for fields in modes with well-defined frequencies, we can obtain relative photon numbers from relative intensities of these fields. A known example being the interference fringes of monochromatic light behind a double-slit which can be calculated from the intensities of electric field strengths and which are proportional to the relative abundancies of photons.

However, the limit from macroscopic numbers of photons (of the order of 10^{18}–10^{20} photons per second in a laser pointer) to single photon sources is far from trivial. The bosonic nature of photons leads to bunching effects which make it difficult to prepare states with definite photon numbers—even photon number $n = 1$ (see, e.g., [24]).

5.4. Wave Optics is Obtained from Geometrical Optics by Geometric Quantization

As has been discussed above (Section 4.2), a formal quantization of the symplectic structure of geometrical optics leads to wave optics. In this sense it is fair to say that wave optics is the quantized theory of geometrical optics, but does "quantized theory" imply that it is a "quantum theory"?

Not necessarily: The fact that the formal mathematical relationship between geometrical optics and wave optics is the same as that between Hamiltonian mechanics and quantum mechanics does not make wave optics a quantum theory. However, wave optics has many of the usual essences or characteristic traits (the German expression "Wesenzug" has been used by [25]) of a quantum theory, such as the superposition principle, interference effects, and the quantization of modes for finite-size systems.

6. Conclusions

The mathematical analogies, discussed throughout the paper, indicate an alternative way to introduce the quantum mechanics of photons using classical models. Whether this method is adequate for schools or only for a university curriculum may be a matter of debate. Hitherto, the phenomenon of detecting single photons as described in the introduction is interpreted in schools as proof of the particle nature of light. However, these phenomena could also be explained by an indivisible ray. The photoelectric effect does not require photons to be point-like: it requires that a quantized amount of energy is transmitted locally to a single atom [2,4], which can also be explained by a ray. The Compton effect might be closer to the scattering of particles, but it can also be explained using rays to which we can assign a wave-length dependent momentum and energy. Moreover, when we refer to the point-like detection of photons on a photographic plate, this is also well explained by a ray which hits this plane.

When talking about "elementary" rays, corresponding to single light-quanta, it might be useful to introduce the notion of rays of finite length. A typical length could be in the range of meters based on the coherence length for monochromatic light. This would account for all events which are observed at "definite" times. It should be noted, however, that the notion of a coherence length is beyond the concepts of classical ray optics.

This alternative approach has some advantages:

1. The comparison of the various limits shown in Figure 1 results in a mathematically consistent and symmetric picture. From a mathematical point of view, one could argue that it makes sense to speak of particles in the case of electrons and of rays in the case of light.
2. In contrast to point-like particles, the concept of a one-dimensional ray circumvents the problem of constructing a three-dimensional probability density from the electromagnetic field (or the potential). It has been argued (see, e.g., [21]) that this is not possible and indeed, as we have indicated above, the relationships between the absolute square of these fields and the (average) number of photons depends on the frequency of the fields.
3. Massive point-like particles have a rest frame and the remaining invariance is the rotation group SO(3). This leads to the usual classification of particles according to their spin. In particular, spin-1 particles have three-dimensional representations for this degree of freedom. Photons however, being massless, have no rest frame and their invariance corresponds to the group SO(2) (for which spin-1 representations are two-dimensional). Rays also have this invariance group of rotations around the ray axis.
4. In particular, in the class room, the analogies between electrons and photons are often emphasized to such an extent that for students they appear to be similar objects [26] and the differences are not sufficiently considered. Different visualizations may prevent such identifications and help to illuminate these differences.

Of course, photons are objects of quantum electrodynamics (QED) and one may argue that classical models are inappropriate to describe quantum mechanical systems anyway. However, there are good reasons not to teach QED in school. Like any model, the notion of a photon as a wave, a particle, or an elementary ray has limitations. However, the notion of a ray allows us to evade the dichotomy of a particle-wave dualism, which in our opinion may be more appropriate for electrons instead for photons. Whether a ray-wave duality or a particle-ray-wave triality is more appropriate depends on the situations for which these models are used.

Author Contributions: Both authors (C.V.F. and T.F.) contributed equally. All authors have read and agreed to the published version of the manuscript.

Funding: This research received no external funding.

Data Availability Statement: Not applicable.

Acknowledgments: We acknowledge stimulating discussions with Andreas Woitzik, Fabian Spallek, and Oliver Passon.

Conflicts of Interest: The authors declare no conflict of interest.

References

1. Passon, O. Was sind Elementarteilchen und gibt es "virtuelle" Photonen? *Prax. Naturwiss.—Phys. Sch.* **2014**, *63*, 44–49. Available online: https://www.researchgate.net/publication/264548826_Was_sind_Elementarteilchen_und_gibt_es_virtuelle_Photonen (accessed on 11 August 2022).
2. Passon, O.; Grebe-Ellis, J. Was ist eigentlich ein Photon? *Prax. Naturwiss.—Phys. Sch.* **2015**, *64*, 46–48. Available online: https://docplayer.org/38337900-Was-ist-eigentlich-ein-photon-1.html (accessed on 11 August 2022).
3. Passon, O. The quasi-history of early quantum theory. *Physics* **2022**, *4*, 880–891. [CrossRef]
4. Passon, O.; Grebe-Ellis, J. Planck's radiation law, the light quantum, and the prehistory of indistinguishability in the teaching of quantum mechanics. *Eur. J. Phys.* **2017**, *38*, 035404. [CrossRef]
5. Fuchs, C.V. Klassische Vorstellungen von Photonen. Master's Thesis, Freiburg University, Freiburg, Germany, 2020. (In German)

6. Bjorken, J.D.; Drell, S.D. *Relativistic Quantum Mechanics*; McGraw-Hill Book Company: New York, NY, USA, 1965. Available online: https://emineter.files.wordpress.com/2018/10/james-d-bjorken-sidney-d-drell-relativistic-quantum-mechanics-1964.pdf (accessed on 11 August 2022).
7. Bogoliubov, N.N.; Shirkov, D.V. *Introduction to the Theory of Quantized Fields*; John Wiley & Sons: Hoboken, NJ, USA, 1959.
8. Berezin, F.A. *The Method of Second Quantization*; Academic Press, Inc. Ltd.: London, UK, 1966. Available online: https://www.sciencedirect.com/bookseries/pure-and-applied-physics/vol/24/suppl/C (accessed on 11 August 2022).
9. Schweber, S.S. *An Introduction to Relativistic Quantum Field Theory*; Harper & Row: New York, NY, USA, 1961.
10. Weinberg, S. *The Quantum Theory of Fields*; Cambridge University Press: Cambridge, UK, 1995; Volumes 1–3. [CrossRef]
11. Hong, C.K.; Ou, Z.Y.; Mandel, L. Measurement of subpicosecond time intervals between two photons by interference. *Phys. Rev. Lett.* **1987**, *59*, 2044–2046. [CrossRef] [PubMed]
12. Grangier, P.; Roger, R.; Aspect, A. Experimental evidence for a photon anticorrelation effect on a beam splitter: A new light on single-photon interferences. *EPL (Euophysr. Lett.)* **1986**, *1*, 173–179. [CrossRef]
13. Loudon, R. Non-classical effects in the statistical properties of light. *Rep. Prog. Phys.* **1980**, *43*, 913–949. [CrossRef]
14. Mandel, L.; Wolf, E.; *Optical Coherence and Quantum Optics*; Cambridge University Press: Cambridge, UK, 1995. [CrossRef]
15. Römer, H. *Theoretical Optics. An Introduction*; Wiley-VCH Verlag GmbH& Co. KGaA: Weinheim, Germany, 2005. Available online: https://www.fulviofrisone.com/attachments/article/411/Romer%20H.%20Theoretical%20optics%20-%20an%20introduction%20 (accessed on 11 August 2022).
16. Bohm, D. *Quantum Theory*; Dover Publications, Inc.: Mineola, New York, NY, USA, 1998.
17. Gottfried, K. *Quantum Mechanics. Volume I: Fundamentals*; CRC Press: Boca Raton, FL, USA, 1989. [CrossRef]
18. Messiah, A. *Quantum Mechanics*; North-Holland Publishing Company: Amsterdam, The Netherlands, 1961. Available online: https://webéducation.com/ebook-quantum-mechanics-by-albert-messiah-pdf/ (accessed on 11 August 2022)
19. Goldstein, H. *Classical Mechanics*; Addison-Wesley Publishing Company, Inc.: Reading, MA, USA, 1950.
20. Guillemin, V.; Sternberg, S. *Symplectic Techniques in Physics*; Cambridge University Press: Cambridge, UK, 1984.
21. Peierls, R. *Surprises in Theoretical Physics*; Princeton University Press: Princeton, NJ, USA, 1979. [CrossRef]
22. Ballentine, L.E. The statistical interpretation of quantum mechanics. *Rev. Mod. Phys.* **1970**, *42*, 358–381 [CrossRef]
23. Jönsson, C. Elektroneninterferenzen an mehreren künstlich hergestellten Feinspalten. *Z. Phys.* **1961**, *161*, 454–474. [CrossRef]
24. Lounis, B.; Orrit, M. Single-photon sources. *Rep. Prog. Phys.* **2005**, *68*, 1129–1179. [CrossRef]
25. Küblbeck, J.; Müller, R. *Wesenszüge der Quantenphysik: Modelle, Bilder and Experimente*; Aulis Verlag Deubner GmbH & Co KG, Köln, Germany, 2002. Available online: https://docplayer.org/40006159-Die-wesenszuege-der-quantenphysik-modelle-bilder-und-experimente.html (accessed on 11 August 2022).
26. Müller, R.; Schecker, H. Schülervorstellungen zur Quanten- und Atomphysik. In *Schülervorstellungen und Physikunterricht*, Schecker, H., Wilhelm, T., Hopf, M., Duit, R., Eds.; Springer Spektrum: Berlin/Heidelberg, Germany, 2018; pp. 209–224. [CrossRef]

physics

Article

Recent Progress on the Sum over Paths Approach in Quantum Mechanics Education

Massimiliano Malgieri [1,*] and Pasquale Onorato [2]

[1] Department of Physics, University of Pavia, Via Bassi 6, 27100 Pavia, Italy
[2] Department of Physics, University of Trento, Via Sommarive 14, Povo, 38123 Trento, Italy
* Correspondence: massimiliano.malgieri@unipv.it

Abstract: In this paper, we present an overview of recent developments in the Feynman sum over paths approach for teaching introductory quantum mechanics to high school students and university undergraduates. A turning point in recent research is identified in the clarification of the distinction between the time-dependent and time-independent approaches, and it is shown how the adoption of the latter has allowed new educational reconstructions to proceed much farther beyond what had previously been achieved. It is argued that sum over paths has now reached full maturity as an educational reconstruction of quantum physics and offers several advantages with respect to other approaches in terms of leading students to develop consistent mental models of quantum phenomena, achieving better conceptual understanding and a higher degree of longitudinal integration of knowledge.

Keywords: quantum physics education; sum over paths; Feynman approach

1. Introduction

The sum over paths approach in physics education originates mainly from two sources: Feynman's path integral formulation of quantum physics [1] and his own divulgation book *QED: the Strange Theory of Light and Matter* [2] (''QED'' stays for Quantum Electro-Dynamics). The latter, in fact, constitutes the first, fundamental sketch of an educational reconstruction of quantum physics based on the path integral formulation. Among the milestones for the development of the approach, one can trace the undergraduate course *Demystifying Quantum Mechanics* on quantum physics held by E.F. Taylor at MIT (Massachusetts Institute of Technology, USA) [3], which had a profound impact in the international physics education research community, and the Advancing Physics project [4] of the British Institute of Physics, an advanced physics course for high schools, designed to attract students to physics, and to give them a good basis for their future progression in the subject at university level, in which J. Ogborn, A. Dobson and collaborators [5] proposed an innovative presentation of quantum physics based on sum over paths. After the turn of the millennium, interest in the sum over paths approach has grown, with several works of great interest, both empirical [6,7] and theoretical [8,9].

Sum over paths has been considered right from the beginning of its history in education as a promising route for teaching the conceptual core of quantum physics to secondary school students and non-physicists. However, some critical points in the approach were also highlighted by some authors (e.g., [10]). These questions and critical remarks can be summarized as follows:

1. Is it possible that using the sum over paths approach may encourage students to retain the classical concept of trajectory, as they may misinterpret Feynman paths as trajectories that are taken alternatively according to some probability rule?
2. The treatment of simple one-dimensional time independent systems may be much more complicated using Feynman's approach than using a standard formulation (i.e., a wavefunction approach).

3. Can the approach be integrated to provide at least an elementary introduction to concepts, related to spin?

2. A summary of Recent Developments

Recent educational research has addressed many of the open issues standing on the sum over paths approach [11], including devising effective educational strategies for discussing time-independent problems such as bound states and tunneling [12,13]; improving the treatment of the uncertainty principle [14]; establishing connections with two state approaches based on spin or light polarization [15]; designing and realizing tools such as interactive simulations and tutorials to sustain students' learning [16,17]. Many of these advances were stimulated or facilitated by the complete clarification of the distinction between a time-dependent and a time-independent sum over paths approach in education [13]. Further improvements included pinpointing and clarifying the educational advantages of sum over paths, including reliable measures of conceptual learning outcomes [18,19] and highlighting the importance of concepts such as path distinguishability, which were not central in the initial educational tests of the approach, but have demonstrated extremely fecund in leading to conceptual understanding of wave particle duality, and allowing modern experimental settings and technologies to be introduced [20].

In the last 10 years, interest in the sum over paths approach has remained high, although research has been led by a few groups, such as the physics education groups at the Universities of Pavia and Trento in Italy and the physics education group at CONICET (Consejo Nacional de Investigaciones Científicas y Técnicas) and the National University of La Plata in Argentina. There has also been related research into the use of Feynman diagrams in education (e.g., Ref. [21]), which can be considered, given the shared underlying philosophy, a "natural" prosecution of quantum instruction in the perspective of sum over paths.

3. Time Dependent vs. Time Independent Sum over Paths Approach

To some degree, the disagreement about the educational usefulness of sum over paths in the last 20 years has been due to a confusion surrounding the role of time in the algorithm for computing detection probabilities. In the 10 years, our group took significant efforts to clarify the situation and show that it is actually possible to use two sum over paths approaches, with a similar general structure but a different identification of physical variables involved in the algorithm of summing over all possible paths. In order to clarify the problem concerning the stationary or time-dependent phenomena, one can analyze the case of the two slit interference of an individual electron.

This problem can be treated in the sum over paths perspective in two ways:

- Considering an initial wavefunction (a "wavepacket") and evolving all the points belonging to it using Feynman's path integral propagator.
- Considering the time-independent problem (at fixed energy) of the propagation of a quantum object from the source to the detector, and using the time-independent propagator (Green function), which basically (apart from a prefactor) reduces to e^{ikx} with x denoting the path length. This approach draws, in addition to Feynman's original works, from more recent research in the area of semiclassical path integrals [22].

The latter approach may be called "stationary" or "time-independent" path integral and the results of the two methods agree, if the energy of the time-dependent wavepacket in the former of the two approaches above is defined with sufficiently small uncertainty (see e.g., Ref. [23]).

In both the following expositions of the time-dependent and time-independent approaches, the concept of action is used. However, secondary school students are rarely exposed to either the full or abbreviated action in the study of classical mechanics. Typically, in both kinds of approaches a rule is given for computing the phase of path amplitudes, and then, if desired, the concept of action is introduced, and the principles of stationary action are derived, after recovering the classical from quantum behavior (correspondence principle) in the short wavelength limit [8].

3.1. The Time-Dependent Sum over Paths Approach

The time-dependent version of the sum over paths approach, derived directly from Feynman's path integral formulation, can be summarized as follows [3,24–26]:

I. The quantum object goes through all possible paths from an initial space–time point (x_i, t_i) to a final space–time point (x_f, t_f).

II. A complex number, often represented by a conventional rotating vector, is associated with each of the paths; its phase angle is proportional to the classical action, $R = \int \mathcal{L}(t)dt$, calculated along the path. Here, $\mathcal{L}$ denotes the Lagrangian.

III. The (normalized) sum of all contributions from the possible paths starting at (x_i, t_i) and ending at (x_f, t_f) gives the time-dependent propagator, which can be understood as the probability amplitude of finding at (x_f, t_f) a quantum object that was initially at (x_i, t_i).

IV. The probability, P, of detecting the quantum object at (x_f, t_f) is then computed by taking the square modulus of the propagator.

Authors have proposed and experimented with several versions of this approach, especially in university education for non-specialists [3,27] and in secondary school [26], but also in introductory courses for first year physics students [28]. These settings are typical of research on sum over paths, as the use of less advanced mathematics is normally in order, and the ability to solve problems may be partly sacrificed in favor of a conceptual understanding of a deep and productive reconstruction of quantum theory. The approach was in general judged promising by physics education researchers although some difficulties were highlighted already from the initial paper by Taylor [3], among which in particular:

- A time-dependent formulation may increase students' confusion about the concepts of quantum paths and classical trajectories.
- The time-dependent treatment obscures the fact that many of the most important predictions of quantum physics are actually time-independent statistics. For example, finding the eigenfunctions for confining potentials usually requires computing the time-dependent propagator and then determining the initial amplitudes that, for the given propagator, are stationary in time, a procedure which appears intricated to students even in the presence of technological aids such as tailored simulation software.

3.2. The Time Independent Sum over Paths Approach

In this approach, the behavior of quantum objects is modelled using a sum-over-paths approach at fixed energy, independent of time. More explicitly, the sum over paths approach at fixed energy, for time-independent problems, can be compactly described as follows:

I. The quantum object goes through all possible paths at fixed energy, E, from an initial point in space, x_i (the source), to a final one, x_f (the detector).

II. A complex number, often represented by a conventional rotating vector, is associated with each of the paths; its phase angle is proportional to the classical abbreviated action, $S = \int p(x)\,dx$, calculated along the path, where $p(x)$ is the particle momentum at point x.

III. The sum of all contributions from the possible paths at fixed energy starting at x_i and ending at x_f gives the energy-dependent propagator, or Green function, which can be understood as the probability amplitude of finding at x_f, independently of arrival time, a particle with defined energy whose source is at x_i.

IV. The probability P of detecting the quantum object at x_f is then proportional to the square modulus of the Green function. For bound systems, the probability is nonvanishing only when the energy E corresponds to an allowed energy level.

In this way, the conceptual structure of Feynman's formulation is entirely preserved, with two main modifications:

- The action R is replaced by the abbreviated action S;
- All paths connecting x_i and x_f, regardless of travel time, are considered.

The "disappearance of time" allows the idea (of educational value in itself) to be introduced that when energy is fixed, time must be completely unknown. This also enables an interesting connection with the time–energy uncertainty to be constructed, which is explored in more detail in Section 5.

The time-independent path integral can be considered a partially different educational reconstruction of quantum mechanics, which in addition to the works of Feynman, also draws from the research on the semiclassical path integral, especially by M.C. Gutzwiller [29] and L.S. Schulman [30]. Within this new perspective, it was possible for researchers to address in a more educationally constructive way several problems of interest in introductory quantum mechanics, such as confined systems with discrete energy levels, as well as the important case of tunnelling [11] with the same elementary mathematical tools used in Feynman's QED [2]. Furthermore, educational advantages in using the time-independent sum over paths approach were found also in the case of open systems: for example, the issue sometimes brought up by students in the treatment of the two slit interference [8], of why it is it that paths of different length are allowed to interfere, since the photon goes through them in different times, is answered right from the beginning in this picture. Paths are independent of time, as they represent the corpuscular equivalent of a plane wave, i.e., a quantum object emitted with infinite uncertainty on emission time, so that paths of different length are actually not distinguishable and the sum of their amplitudes must be considered. The focus, right on the beginning, on the importance of path indistinguishability for producing interference (see Section 7) allows students to construct a more consistent idea of wave particle duality. In the following Section, we provide some recent results deriving from the adoption of a time-independent sum over paths approach.

4. Treatment of Stationary Problems

4.1. Infinite Square Well

The ''particle in a box" problem is the paradigmatic example for the treatment of bound systems in the sum over paths approach. A quantum object is confined in a square potential well with infinite depth and width L. For a fixed energy E, the particle can reach the detector x_f starting from a source at x_i through one of four families of paths, depicted in Figure 1. The phasor corresponding to each path is computed by the usual rules, with each reflection contributing a π phase loss, and the result is that amplitudes associated with all possible paths interfere constructively when the value of energy corresponds to an allowed energy level (Figure 2; the computed amplitude sum is in the right window) and destructively otherwise.

Analytically, the allowed energy levels (the poles of the energy dependent Green function) can be determined uniquely from the condition that two paths differing for a full back and forth round trip in the well are in phase. Thus, the full Green function can be determined and the stationary wave functions (eigenfunctions) can be also evaluated. The set of GeoGebra tutorials [31], created to support student understanding of quantum concepts within the sum over paths approach [16,17], have been tested with both secondary school students and pre-service and in-service teachers in several courses. Since the simulations are available for free on the GeoGebra website [31], they have also occasionally been modified by teachers to produce simulations of different physical situations or more suitable to the needs [32], or considered in other research [33].

In a similar fashion to the square well potential, other confined systems of interest (e.g., the harmonic oscillator, the finite square well [12], the particle confined on a circumference [14]) can be discussed with students in a conceptually consistent way, using the same simple mathematical tools (vector amplitudes) used in the treatment of open systems, such as the two slit interference.

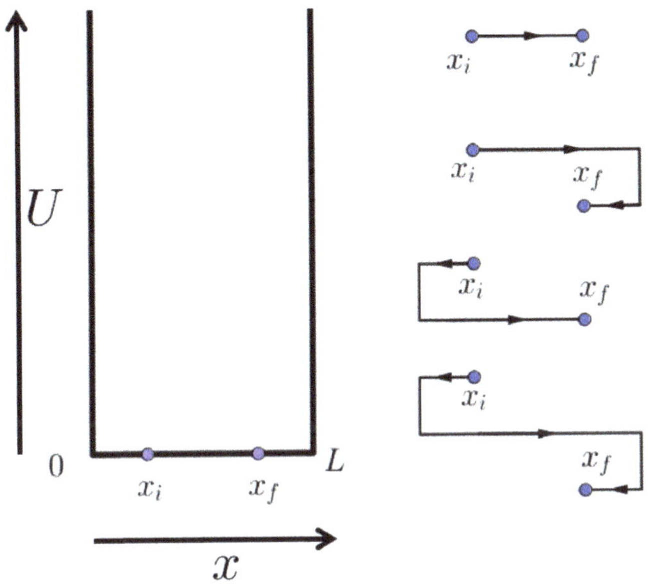

Figure 1. Left: a representation of the one-dimensional infinite square potential well with two points arbitrarily taken as initial, x_i, and final, x_f. As common in these representation, the vertical axis U is an energy scale. **Right:** the four basic possible routes exist from the source to the detector, including possible flections on the potential walls. Theoretically, all the paths that can be constructed by adding to any of the above an arbitrary number of full back and forth routes should be considered. Note that each reflection from a wall brings a $e^{-i\pi}$ contribution (inverts the sign) of the amplitude, associated to that path.

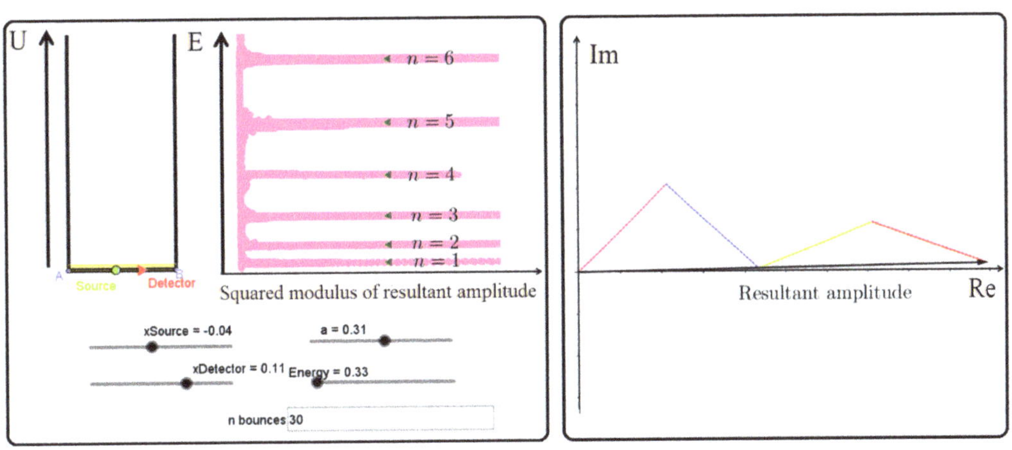

Figure 2. Partial snapshot of the simulation of the infinite square potential well. **Left:** a drawing of the potential, the initial and final points, and the numerically computed energy levels for $n = 1$ to 6. The sliders of the GeoGebra simulation are shown as grey lines with the input values shown by solid black circles. a is the halfwidth. In this particular case, the value of Energy corresponds to a stationary state (energy level) for the given value of a. The values of xSource and xDetector correspond to x_i and x_f in the text and in this case neither of them is placed on a wavefunction node, see below for details.

Right: detail of the imaginary part vs. real part of the amplitude computation for a particular value of energy, in this case, very close to one of the energy eigenvalues. Note that each one of the colored segments is actually composed of $N = 30$ parallel tiny arrows placed head to tail. In fact, in this condition, the amplitudes for paths belonging to the same one of the families shown in the right part of Figure 1 are in phase. The angle between amplitudes corresponding to different path families (the angle between colored segments, in the right of Figure 2) depends on the position of x_i and x_f, and may lead to destructive interference if either point coincides with a wavefunction node, as expected, since for stationary wavefunctions, the amplitude only vanishes at nodes. A special case is the limit of energy $E \to 0$, for which the paths of a given family are in phase, but due to the even or odd number of path inversions for different families, the resultant of the sum of all four families is identically zero for all x_i and x_f. For other values of energy, amplitudes within each individual family of paths will interfere destructively and in the limit of the number of paths $n \to \infty$ the resultant amplitude will vanish.

Despite the conceptual simplicity of the treatment of one-dimensional systems, it has to be remembered that there have been technical difficulties in extending path integral techniques to radial coordinates beyond the simplest cases [30,34], and progress in obtaining path integral solutions to three dimensional problems in radial coordinates, including the hydrogen atom, has been slow [35,36]. Correspondingly, a sum over paths approach may not be the most suitable way to deal with these problems, which typically appear at advanced undergraduate and graduate levels.

4.2. Tunneling from a Square Barrier

In the context of the time-independent sum over paths approach, the problem of tunneling from a square barrier can also be solved analytically, and the main conceptual elements of the solution can be shown to students through a simulation (Figure 3). Transmission and reflection coefficients at the barrier borders are computed by the simulation through the equivalent for massive particles of Fresnel coefficients [12], but like in the case of the square well potential, important features of the solution, such as the formation of energy-dependent resonances in the transmitted amplitude, are due to constructive interference between paths, which undergo multiple reflections within the barrier.

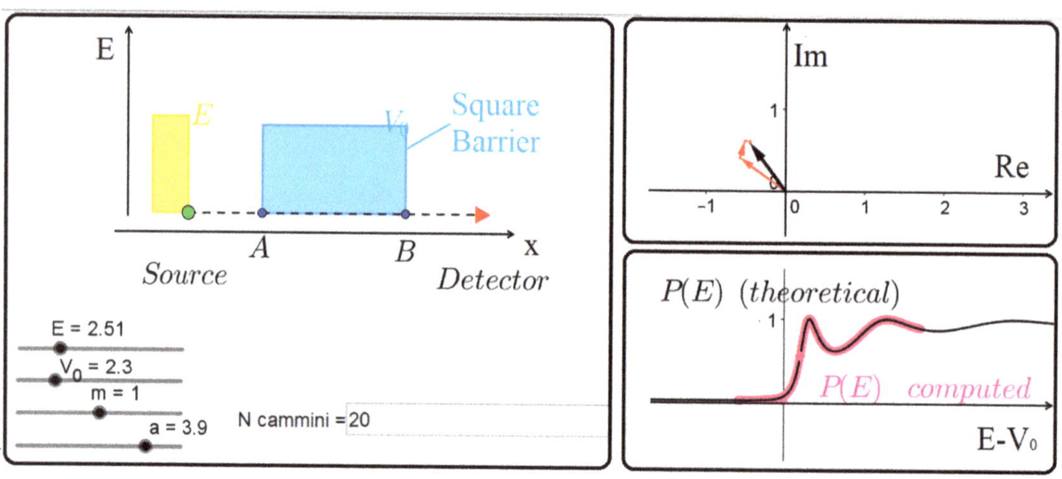

Figure 3. Partial snapshot of the simulation of square barrier tunneling for a massive particle in the framework of the time-independent sum over paths approach. **Left**: the green dot represents the source and the red triangle the detector. The yellow rectangle represents the incoming particle energy, as compared to the barrier energy (light blue). m is the particle mass in natural atomic units (so $m = 1$

is the electron mass). The word 'cammini' stands for 'paths' in Italian. **Top right**: the representation of the sum of amplitudes (in this case the amplitudes of paths reflecting inside the barrier have progressively lower absolute value as the number of internal reflections increases). The red arrows indicate the amplitudes associated to individual paths, which are summed to form the final amplitude (black arrow). **Bottom right**: the probability, $P(E)$, of revealing the quantum object at the detector beyond the barrier (transmission probability) as a function of the difference between the particle energy and the barrier height, compared to the theoretical formula.

In the typical structure of a teaching-learning sequence based on the sum over paths approach (e.g., Ref. [18]), the topic of tunneling is treated before bound systems and energy quantization, with the case of the resonant scattering of a photon between two semi-reflecting mirrors (Figure 4) playing the role of a transition case between open and bound systems.

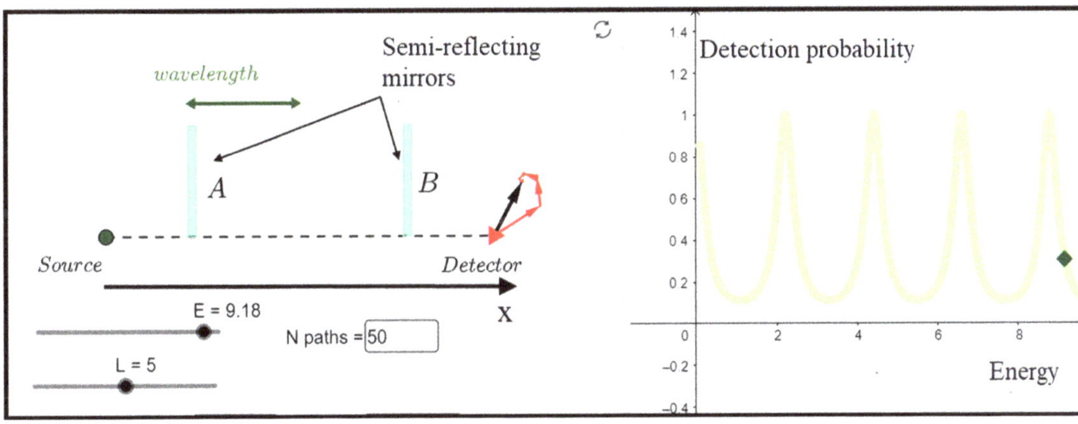

Figure 4. Partial snapshot of the simulation of resonant scattering between two semi-reflecting mirrors for a photon in the framework of the time-independent sum over paths approach. **Left:** the setup with the green dot representing the source and the red triangle the detector. The green arrow represents the wavelength of the photon directed towards a system of two semi-reflecting mirrors. **Right:** the probability of detecting the photon at the detector beyond the mirrors as a function of energy, displaying the typical equispaced resonances. The diamond shows the energy used in the simulation. The red arrows indicate the amplitudes associated to individual paths, which are summed to form the final amplitude (black arrow).

5. Stationary Sum over Paths Approach and the Time-Energy Uncertainty Relationship

In the fixed-energy sum over paths approach, the uncertainty in the travel time of the quantum object from source to detector is considered infinite. Thus, for bound systems, infinitely many paths, going through an arbitrary number of back and forth roundabouts within the confining potential, need to be considered in order to obtain the allowed energy levels. On the other hand, it was shown recently [14] that, if the assumption of infinite uncertainty in travel time is weakened, considering only a finite, though large, uncertainty in time, it is possible to derive from the time-independent sum over paths approach, through geometrical considerations and simple algebra, a time-energy indeterminacy relationship of the form $\Delta E \cdot \Delta t \approx \hbar$, where $\hbar$ is the reduced Planck's constant. In fact, if only a finite number of paths are considered, namely those whose travel times differ by less than the time uncertainty, the resulting approximate Green function will not have sharp divergences, but widened peaks, whose width in energy can be described such relationship.

The derivation is not valid only for the infinite square well shown in Figure 5, but for any confined system (in Ref. [14] the particle on a ring is treated). The time indeterminacy,

Δt, thus derived can be given different interpretations [14]: (a) upper limit on the time the quantum object can have spent in the confined system before measurement, which is the most literal interpretation within the sum over paths approach: (b) coherence time for paths of different length, i.e., timescale over which they can still considered indistinguishable; and (c) lifetime of the quantum state, due to external, unspecified reasons, which make it an unstable state. Among the interpretations proposed by the authors, the third one seems the most promising in education, given the relevance of the lifetime–linewidth relationship in the elementary treatment of the time-energy uncertainty principle. Thus, it may be worth explaining the terms of such interpretation in some more detail. The lifetime–linewidth relationship, originally derived by G. Gamow in the context of alpha decay [37] but valid in a wide range of different contexts, applies to quasi-bound states, which are resonances in the continuous spectrum of object temporarily confined by a potential barrier, which nonetheless possess a positive total energy. Gamow's derivation made use of a semiclassical picture, in which the quantum object, bounced within the walls of the potential an indeterminate number of times, each time with a finite probability of escaping the well. In this picture, the time indeterminacy Δt is the expected value of the time the quantum object remains in the confined state. Thus, the derivation proposed in Ref. [14] can be seen as a simplified (at a level accessible also to advanced high school students) treatment of Gamow's analysis, in which rather than assigning a finite probability to the escape of the quantum object from the potential at each bounce, a hard limit is imposed on the dwell time. Note that a rigorous analysis of Alpha decay in terms of the semiclassical path integral (which involves assigning progressively decreasing amplitudes to paths) was given in Ref. [38], where its results are compared to those of Gamow's original treatment.

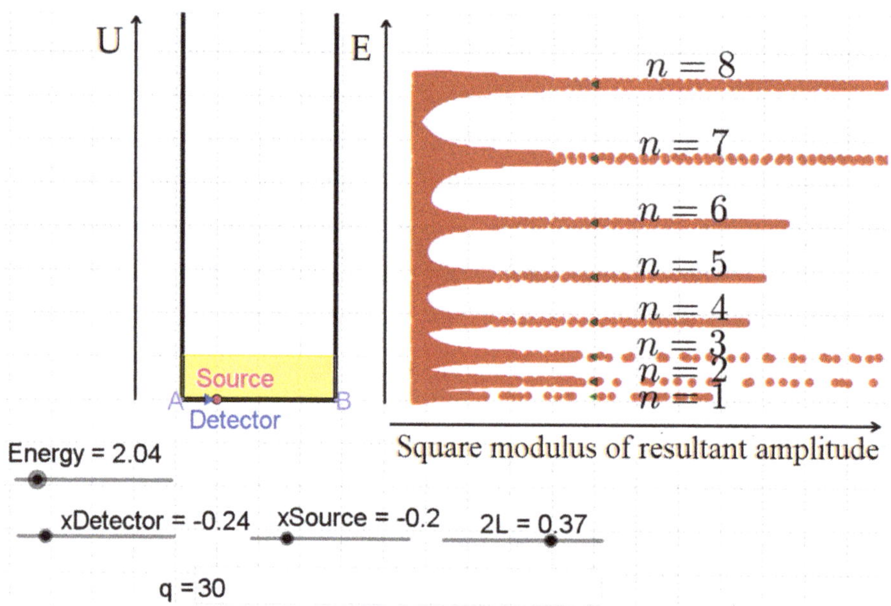

Figure 5. Graph of the approximate spectrum of the particle-in-a-box system when only a finite number of paths (here, $N = 120$ for all values of energy) is considered. The red dot represents the source, the green triangle the detector. The spread of spectral lines corresponding to energy levels into Lorentzian-like shapes is evident. The width of the peaks is not uniform because in this simulation N is kept fixed for all values of energy rather than depending on a fixed time uncertainty, Δt, and the particle momentum, p.

6. Connections with Spin: From the Hong-Ou-Mandel Experiment to Quantum Computing

The Hong-Ou-Mandel (HOM) experiment [39] demonstrates interference between indistinguishable processes, and its generalized version [15] using electrons can be used to explain the properties of bosons and fermions. The general setup of the HOM experiment is depicted in Figure 6. The experiment is known in its original version performed with photons, as it demonstrates the possibility of two photon interference due to perfect time overlap and consequent indistinguishability. A less known version of the experiment with electrons was also performed [40].

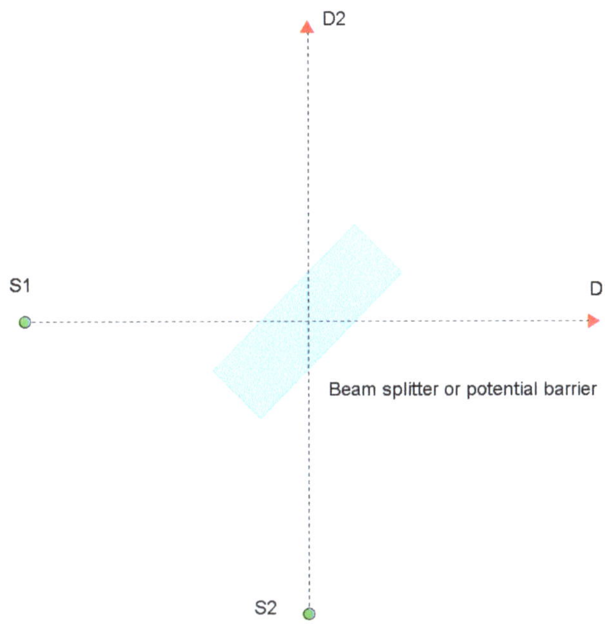

Figure 6. Setup of a generalized HOM (Hong-Ou-Mandel) experiment. Two sources (S1 and S2) emit simultaneously indistinguishable quantum objects (either photons or electrons), which are directed towards either a 50% beam splitter, or in the case of electrons, a potential barrier tailored to have 50% reflection and 50% transmission probability for the specific energy of the incoming electrons. The quantum objects may then be revealed at detectors D1 or D2.

Discussing this experiment with students involves first of all a change of perspective, from performing the sum over all possible paths, to performing a sum over all possible processes or histories for a given setup. The crucial step in such direction is to explain that, for processes involving more than one quantum object, amplitudes of histories can be derived from the amplitudes of single particles paths through a multiplicative operation (mirroring the multiplication between complex numbers): the result, in the language of Feynman's QED, is an arrow whose length is the product of the lengths of the single-particle arrows, and whose phase is the sum of their phases [2]. At this point, the three possible distinct outcomes of the experiment (two particles found at D1, two particles at D2, or one particle in each, referring to Figure 6) can each be represented by two diagrams differing by the exchange of identical particles (see Figure 3 in Ref. [15] and the text there). Next, the outcomes of the real experiment are considered. With perfect time coherence, and so perfect indistinguishability of the two quantum objects, photons are always detected either both at D1, or both at D2. On the other hand, in the same conditions, electrons are always found at different detectors. Considering also the phase shift properties of beam splitters, the only way to solve such a puzzle is to postulate different exchange rules for the two types of quantum objects: for indistinguishable fermions, a change of sign for the global amplitude is involved for diagrams, which only differ by the exchange of two particles; while for indistinguishable bosons, no such change of sign is involved. The possibility to introduce, starting from the phenomenology of outcomes of modern experiments, a key element of the spin-statistics connection, namely the different rules related to indistinguishability for bosons and fermions, on the one hand reinforces the centrality of the concept of indistinguishability in quantum mechanics (which is a key idea of the HOM experiment), and on the other hand provides a first connection of the sum over paths approach with topics related to spin.

Furthermore, there is currently ongoing research on the integration of sum over paths with two state approaches [41], especially in the context of teaching introductory elements of quantum computation and communication. In fact, in this context, in order to show students applications of quantum computing that are not entirely trivial (e.g., the Deutsch and Grover algorithms), it is necessary to code more than just a single qubit (often at least two). The educational issue in choosing, as the physical realization of such qubits, different quantum objects (e.g., two or more electrons) is that by doing so the quantum gates become very difficult to interpret as physical systems within the reach of secondary school students. The consequence is that at that point the treatment of quantum algorithms becomes completely symbolic and algebraic, disconnected by any concrete physical representation. One alternative that is being explored [42], which allows two qubit quantum logical circuits using simple optical devices to be constructed, is to use photons and code one qubit as the polarization state, the second qubit as the choice between possible paths (typically at a beam splitter). Of course, for the second qubit, all the rules typically introduced in sum over paths concerning, for example, phase displacements at reflection and transmission are valid. This approach can be considered a prosecution of the one pioneered by Daniel F. Styer in his book *The Strange World of Quantum Mechanics* [27].

7. Path Indistinguishability and the Zhou–Wang–Mandel Experiment

The HOM experiment is only one example of the fecundity of the concept of path distinguishability/indistinguishability in education [20]. Another experimental result that has proven highly useful in education is the Zhou–Wang–Mandel (ZWM) experiment [43], which highlights in a peculiarly sharp way the loss of interference resulting from rendering quantum processes experimentally distinguishable. The ZWM experiment is conceptually a concrete realization of a two slit interference experiment with which-way detectors; but which way information is obtained not by physical detection of the photon passing through a slit, but by detection of a secondary photon that is emitted by a non-linear crystal at the same time as the primary one. Discussion of this experiment is highly instructive for students, as they are introduced to a case in which interference can be made to appear or disappear, without physically touching the quantum object, which is (supposedly) only responsible for the appearance of interference fringes, i.e., the "signal" photon arriving at the primary detector. While the initial reaction of many students is to declare that the experiment works by way of magic, a more thorough discussion leads to a deeper understanding of the modern meaning of wave particle duality in quantum mechanics as a relationship of complementarity between interference fringe visibility and path distinguishability [44].

8. Educational Outcomes

Several tests of the educational effectiveness of the sum over paths approach were performed in recent years. In Ref. [18], based on an experimentation in a class of final year secondary school students, the authors found that Feynman's approach could lead students to build consistent, detailed, and integrated mental models of wave particle duality, as tested within Knowledge Integration theory [45]. Furthermore, the approach could also help overcome some other integration issues in teaching–learning quantum mechanics, such as students building completely different mental models for the photon and the electron. In this study, specific strategies are adopted in order to reduce the possibility that students interpret paths as classical trajectories. They emphasized that the sum over possible paths has to be considered a representation of the mathematical model of quantum theory, and not of the actual physical reality. Moreover, students were guided through a detailed analysis, in the context of the two slit and Mach Zehnder settings, of the inconsistency of considering paths as mutually exclusive classical trajectories (e.g., that the quantum object passed through one or the other slit, or one or the other arm of the interferometer). In the teaching experiment of Ref. [18], the issue of students appearing to reason as if paths were alternative classical trajectories was limited to one student. The incidence of other deterministic and hybrid conceptions was also reportedly low or non-existent. Advantages in terms of consistence between students' mental models of the photon and electron were also reported in Ref. [19], which discusses an experimentation in a class of 16–17-year-old students. In Ref. [46], which is based on an experimentation with four classes of 16–17-year-old secondary school students, the authors report advantages in helping students form consistent and integrated models of light, by connecting in a unified perspective phenomena related to ray optics, wave optics, and the quantum theory of light. These results, which echo the findings reported in Ref. [47], appear especially important in view of the extremely fragmented character of the topic of light within the secondary school curriculum in many countries. Convincing results were obtained in the context of teacher education in several studies [16,17] and it was highlighted that the deep conceptual understanding provided by the sum over paths approach in in-service courses can render a teacher confident enough to undertake the enterprise of treating quantum physics in the classroom, even in contexts (such as the Italian system) in which it is possible, and quite frequent, that secondary school physics teachers have no formal instruction on the topic at all (for example, having a degree in mathematics or engineering).

9. Discussion and Recapitulation of Educational Perspectives

Based on the results of research literature, and several years of direct experience with using the sum over paths approach in teacher education, let us summarize the main educational advantages offered by in the following way:

1. On the mathematical level, the sum over paths approach allows quantum phenomena to be discussed using quite a simple formal language. At its heart, such a possibility is due to the fact that, rather than finding solutions to the Schrödinger equation, Feynman's method constructs the Green function for the same equation, representing it as a sum of complex amplitudes computed over all possible paths. In educational practice, complex amplitudes associated with paths can be represented and added up as vectors or "little arrows", a strategy directly derived from the one used by Feynman himself, which greatly reduces the stress on student's cognitive resources while learning the basics of quantum theory. The recent advances in the design of teaching–learning sequences based on sum over paths, mostly due to a significant clarification of the subject matter and the adoption of the time-independent version of the approach, allow the same simple mathematical machinery essentially to be treated with all problems, which are typically solved with the one-dimensional time-independent Schrodinger equation. Thus, sum over paths can be considered an attractive option not only for secondary education, but also for the introduction of elements of quantum physics to non-physicists.
2. On the conceptual level, sum over paths has the unique peculiarity of offering students a clear and unambiguous representation of one of the most profound quantum mysteries, namely wave particle duality. There are two basic ingredients that contribute to forming such conceptual understanding. The first one consists of the distinction between classical and quantum ways of computing probabilities, which is at the heart of the approach, and allows what is always "corpuscular" in quantum objects' behaviour (they are always revealed as discrete entities at detectors) and what may, or may not,

be "wave-like" (the statistics of their detection events) to be clearly distinguished. The second ingredient is the focus on path distinguishability/indistinguishability, which allows a modern understanding of duality to be constructed in which an either/or (particle or wave) idea of the quantum object is replaced by a continuum of wave-like and particle-like behaviours, regulated by the respective weight of path distinguishability and fringe visibility. Furthermore, while going through this educational path, students generalize the concept of 'sum over paths' to 'sum over histories', and by doing so they construct a language capable to discuss modern experiments and technologies based on quantum optics, and in principle, to understand the conceptual meaning of simpler Feynman diagrams [15]. Finally, modern educational reconstructions based on sum over paths can offer deep insight into the origin of energy quantization for bound systems, and help clarify the meaning of the time-energy uncertainty principle.

3. At the level of knowledge integration, the sum over paths formulation can make the classical limit (correspondence principle) completely transparent [8], and provide a unifying perspective on the nature of light, connecting ray optics, wave optics, and the quantum behaviour of photons. The approach allows students to build consistent mental models for photons and electrons, in which differences (the dispersion relation, the exchange rule) are highlighted that build on a common basic model of the quantum object. Rather than presenting quantum theory as a set of disconnected formulas for different situations, as it appears in many secondary school textbooks, in sum over paths the subject matter is presented as an organic theory, with the additional advantage of offering the possibility to teachers to develop their own exercises and problems, applying the computational rules to new situations.

10. Conclusions

In this overview of research on the sum over paths approach for teaching introductory quantum physics, we have argued that such an approach, whose history started in the late 1980s, has reached full maturity in the second decade of the XXI century. Research has addressed most of the critical points enumerated in Section 1, and while there may not be a consensus as to whether they have been fully resolved (especially concerning the first of the three issues, on which data, while encouraging, is rather scarce), progress has been significant, and can be considered decisive especially concerning the treatment of bound systems. Furthermore, new experimentations have reinforced evidence on the educational advantages of sum over paths, demonstrating that the approach can help researchers and educators improve educational outcomes in terms of conceptual understanding and knowledge integration, and be an invaluable aid in the introduction of quantum technologies, an issue which is increasingly felt as central and urgent.

Author Contributions: Conceptualization, M.M. and P.O.; methodology, M.M. and P.O.; writing—original draft preparation, M.M.; writing—review and editing, M.M. and P.O. All authors have read and agreed to the published version of the manuscript.

Funding: This research received no external funding.

Data Availability Statement: No data were generated or used in this work.

Conflicts of Interest: The authors declare no conflict of interest.

References

1. Feynman, R.P. Space-time approach to non-relativistic quantum mechanics. *Rev. Mod. Phys.* **1948**, *20*, 367–387. [CrossRef]
2. Feynman, R.P. *QED: The Strange Theory of Light and Matter*; Princeton University Press: Princeton, NJ, USA, 1985.
3. Taylor, E.F.; Vokos, S.; O'Meara, J.M.; Thornber, N.S. Teaching Feynman's sum-over-paths quantum theory. *Comput. Phys.* **1998**, *12*, 190–199. [CrossRef]
4. Ogborn, J.; Whitehouse, M. (Eds.) *Advancing Physics AS*; Institute of Physics Publishing: Bristol, UK, 2000.
5. Dobson, K.; Lawrence, I.; Britton, P. The A to B of quantum physics. *Phys. Educ.* **2000**, *35*, 400–405. [CrossRef]

6. Hanc, J.; Tuleja, S. The Feynman Quantum Mechanics with the help of Java applets and physlets in Slovakia. In Proceedings of the 10th Workshop on Multimedia in Physics Teaching and Learning (MPTL), Berlin, Germany, 5–7 October 2005; Available online: https://citeseerx.ist.psu.edu/viewdoc/download?doi=10.1.1.468.3374&rep=rep1&type=pdf (accessed on 15 August 2022).
7. de los Angeles Fanaro, M.; Otero, M.R.; Arlego, M. Teaching basic quantum mechanics in secondary school using concepts of Feynman path integrals method. *Phys. Teach.* **2012**, *50*, 156–158. [CrossRef]
8. Ogborn, J.; Taylor, E.F. Quantum physics explains Newton's laws of motion. *Phys. Educ.* **2005**, *40*, 26–34. [CrossRef]
9. Hanc, J. The time-independent Schrodinger equation in the frame of Feynman's version of quantum mechanics. In Proceedings of the 11th Workshop on Multimedia in Physics Teaching and Learning, Szeged, Hungary, 20–22 September 2006; Available online: https://physedu.science.upjs.sk/modelovanie/files/hanc_szeged_2006.pdf (accessed on 15 August 2022).
10. Stefanel, A. Impostazioni e percorsi per l'insegnamento della meccanica quantistica nella scuola secondaria. *Giorn. Fis.* **2008**, *49*, 15–53. [CrossRef]
11. Malgieri, M. State of the Art of the Sum over Paths Approach in Education. In Proceedings of the 3rd World Congress on Physics Education: Innovating Physics: From Research to Practice (WCPE III 2021), Hanoi, Vietnam, 13–17 December 2021; Available online: https://indico.cern.ch/event/1025696/contributions/4630286/ (accessed on 15 August 2022).
12. Malgieri, M.; Onorato, P.; De Ambrosis, A. A sum-over-paths approach to one-dimensional time-independent quantum systems. *Am. J. Phys.* **2016**, *84*, 678–689. [CrossRef]
13. Malgieri, M.; Onorato, P. Educational reconstructions of quantum physics using the sum over paths approach with energy dependent propagators. *J. Phys. Conf. Ser.* **2021**, *1929*, 012047. [CrossRef]
14. Malgieri, M.; Zuccarini, G.; Onorato, P. Approximate time–energy uncertainty relationship from the fixed-energy sum over paths approach. *Canad. J. Phys.* **2022**. [CrossRef]
15. Malgieri, M.; Tenni, A.; Onorato, P.; De Ambrosis, A. What Feynman could not yet use: The generalised Hong–Ou–Mandel experiment to improve the QED explanation of the Pauli exclusion principle. *Phys. Educ.* **2016**, *51*, 055002. [CrossRef]
16. Malgieri, M.; Onorato, P.; De Ambrosis, A. Teaching quantum physics by the sum over paths approach and GeoGebra simulations. *Eur. J. Phys.* **2014**, *35*, 055024. [CrossRef]
17. Malgieri, M.; Onorato, P.; De Ambrosis, A. GeoGebra simulations for Feynman's sum over paths approach. *Il Nuovo Cim. C* **2018**, *41*, 124. [CrossRef]
18. Malgieri, M.; Onorato, P.; De Ambrosis, A. Test on the effectiveness of the sum over paths approach in favoring the construction of an integrated knowledge of quantum physics in high school. *Phys. Rev. Phys. Educ. Res.* **2017**, *13*, 010101. [CrossRef]
19. de los Angeles Fanaro, M.; Arlego, M.; Otero, M.R. A didactic proposed for teaching the concepts of electrons and light in secondary school using Feynman's path sum method. *Eur. J. Phys. Educ.* **2017**, *3*, 1–11. Available online: http://www.eu-journal.org/index.php/EJPE/article/view/107 (accessed on 7 July 2022).
20. Hochrainer, A.; Lahiri, M.; Erhard, M.; Krenn, M.; Zeilinger, A. Quantum indistinguishability by path identity and with undetected photons. *Rev. Mod. Phys.* **2022**, *94*, 025007. [CrossRef]
21. Passon, O.; Zügge, T.; Grebe-Ellis, J.; Passon, O.; Zügge, T.; Grebe-Ellis, J. Pitfalls in the teaching of elementary particle physics. *Phys. Educ.* **2018**, *54*, 015014. [CrossRef]
22. Blumel, R. *Advanced Quantum Mechanics: The Classical-Quantum Connection*; Jones & Bartlett Publishers: Sudbury, MA, USA, 2010.
23. Sawant, R.; Samuel, J.; Sinha, A.; Sinha, S.; Sinha, U. Nonclassical paths in quantum interference experiments. *Phys. Rev. Lett.* **2014**, *113*, 120406. [CrossRef] [PubMed]
24. Hanc, J.; Taylor, E.F. From conservation of energy to the principle of least action: A story line. *Am. J. Phys.* **2004**, *72*, 514–521. [CrossRef]
25. de los Ángeles Fanaro, M.; Otero, M.R. Basics Quantum Mechanics teaching in secondary school: One conceptual structure based on Paths Integrals Method. *Lat. Am. J. Phys. Educ.* **2008**, *2*, 103–112. Available online: http://www.lajpe.org/may08/03_Maria_Fanaro.pdf (accessed on 7 July 2022).
26. Cuppari, A.; Rinaudo, G.; Robutti, O.; Violino, P. Gradual introduction of some aspects of quantum mechanics in a high school curriculum. *Phys. Educ.* **1997**, *32*, 302–308. [CrossRef]
27. Styer, D.F. *The Strange World of Quantum Mechanics*; Cambridge University Press.: Cambridge, UK, 2000. [CrossRef]
28. McGinness, L.P.; Savage, C.M. Action physics. *Am. J. Phys.* **2016**, *84*, 704–708. [CrossRef]
29. Gutzwiller, M.C. Phase-Integral Approximation in Momentum Space and the Bound States of an Atom. *J. Math. Phys.* **1967**, *8*, 1979–2000. [CrossRef]
30. Schulman, L.S. *Techniques and Applications of Path Integration*; John Wiley and Sons: New York, NY, USA, 1981; Available online: https://vdoc.pub/documents/techniques-and-applications-of-path-integration-gn5q7aenmls0 (accessed on 7 July 2022).
31. GeoGebra. Available online: https://www.geogebra.org/m/Q6waMV2v (accessed on 7 July 2022).
32. Mastropaolo, G. "Cammini di Feynman" in Laboratorio di G. Mastropaolo. Available online: https://mastropaolo.net/didattica-con-scratch/fisica/cammini-di-feynman (accessed on 7 July 2022).
33. Solvang, L.; Haglund, J. How can GeoGebra support physics education in upper-secondary school—A review. *Phys. Educ.* **2021**, *56*, 055011. [CrossRef]
34. Kleinert, H. *Path Integrals in Quantum Mechanics, Statistics, Polymer Physics, and Financial Markets*; World Scientific: Singapore, 2009. [CrossRef]
35. Peak, D.; Inomata, A. Summation over Feynman histories in polar coordinates. *J. Math. Phys.* **1969**, *10*, 1422–1428. [CrossRef]

36. Duru, I.H.; Kleinert, H. Quantum Mechanics of H-Atom from Path Integrals. *Fortschr. Phys.* **1982**, *30*, 401–435. [CrossRef]
37. Gamow, G. The quantum theory of nuclear disintegration. *Nature* **1928**, *122*, 805. [CrossRef]
38. Holstein, B.R. Understanding alpha decay. *Am. J. Phys.* **1996**, *64*, 1061–1071. [CrossRef]
39. Hong, C.K.; Ou, Z.Y.; Mandel, L. Measurement of subpicosecond time intervals between two photons by interference. *Phys. Rev. Lett.* **1987**, *59*, 2044–2046. [CrossRef]
40. Lim, Y.L.; Beige, A. Generalized Hong–Ou–Mandel experiments with bosons and fermions. *New J. Phys.* **2005**, *7*, 155. [CrossRef]
41. Maries, A.; Sayer, R.; Singh, C. Can students apply the concept of "which-path" information learned in the context of Mach–Zehnder interferometer to the double-slit experiment? *Am. J. Phys.* **2020**, *88*, 542–550. [CrossRef]
42. Sutrini, C.; Malgieri, M.; Macchiavello, C. Quantum technologies: A course for teacher professional development. *J. Phys. Conf. Ser.* **2022**, *2297*, 012018. [CrossRef]
43. Zhou, X.Y.; Wang, L.J.; Mandel, L. Induced coherence and indistinguishability in optical interference. *Phys. Rev. Lett.* **1991**, *67*, 318–321. [CrossRef] [PubMed]
44. Englert, B.G. Fringe visibility and which-way information: An inequality. *Phys. Rev. Lett.* **1996**, *77*, 2154–2157. [CrossRef] [PubMed]
45. Linn, M.C. Designing the knowledge integration environment. *Int. J. Sci. Educ.* **2000**, *22*, 781–796. [CrossRef]
46. Fanaro, M.; Otero, M.; Elgue, M. Implementation of a proposal to teach quantum mechanics concepts from the Multiple Paths of Feynman applied to the light. In Proceedings of the GIREP-MPTL International Conference on Teaching/Learning Physics: Integrating Research into Practice, Palermo, Italy, 7–12 July 2014; pp. 225–232. Available online: https://www.researchgate.net/profile/Maria-Otero-6/publication/262186885_Implementation_of_a_proposal_to_teach_quantum_mechanics_concepts_from_the_Multiple_Paths_of_Feynman_applied_to_the_light/links/55c7dd6208aeca747d667def/Implementation-of-a-proposal-to-teach-quantum-mechanics-concepts-from-the-Multiple-Paths-of-Feynman-applied-to-the-light.pdf (accessed on 7 July 2022).
47. Sutrini, C.; Malgieri, M.; De Ambrosis, A. Bubble: Experimenting with Feynman's sum over paths approach in the secondary school. *Il Nuovo Cim. C* **2019**, *42*, 1–10. [CrossRef]

physics

MDPI

Article

Assessing Engineering Students' Conceptual Understanding of Introductory Quantum Optics

Philipp Bitzenbauer [1,*], Joaquin M. Veith [2], Boris Girnat [2] and Jan-Peter Meyn [1]

1 Professur für Didaktik der Physik, Friedrich-Alexander-Universität Erlangen-Nürnberg, Staudtstr. 7, 91058 Erlangen, Germany
2 Institut für Mathematik und Angewandte Informatik, Stiftung Universität Hildesheim, Samelsonplatz 1, 31141 Hildesheim, Germany
* Correspondence: philipp.bitzenbauer@fau.de

Abstract: Quantum technologies have outgrown mere fundamental research in laboratories over recent years, and will facilitate more and more potentially disruptive applications in a wide range of fields in the future. In foresight, qualification opportunities need to be implemented in order to train qualified specialists, referred to as the future quantum workforce, in various fields. Universities world-wide have launched qualification programmes for engineers focusing on quantum optics and photonics. In many of these programmes, students attend courses on quantum physics contextualized via quantum optics experiments with heralded photons, because: (1) their experimental and physical foundations may be directly leveraged to teaching a number of quantum technology applications, and (2) physics education research has provided empirical evidence, according to which such quantum optics-based approaches are conducive to learning about quantum concepts. While many teachers are confident about the effectiveness of their concepts, there is little empirical evidence due to the lack of content-area-specific research tools. We present a 16-item concept inventory to assess students' conceptual understanding of quantum optics concepts in the context of experiments with heralded photons adopted from a test instrument published in the literature. We have administered this Quantum Optics Concept Inventory as a post-test to $N = 216$ students after instruction on quantum optics as part of an undergraduate engineering course. We evaluated the instruments' psychometric quality, both in terms of classical test theory, and using a Rasch scaling approach. The Quantum Optics Concept Inventory enables a reliable measure ($\alpha = 0.74$), and the data gathered show a good fit to the Rasch model. The students' scores suggest that fundamental quantum effects pose striking learning hurdles to the engineering students. In contrast, most of the students are able to cope with the experimental and technical foundations of quantum optics experiments with heralded photons and their underlying principles, such as the coincidence technique used for the preparation of single-photon states. These findings are in accordance with prior research, and hence, the Quantum Optics Concept Inventory may serve as a fruitful starting point for future empirical research with regard to the education of the future quantum workforce.

Keywords: quantum optics; concept inventory; Rasch scaling; quantum technology

1. Introduction

Quantum technologies have evolved dynamically in recent years [1]: From purely fundamental research in laboratories, they have increasingly moved toward potential disruptive applications in a wide range of fields until today [2,3]. To establish quantum technologies in future industry sustainably, it becomes necessary to train qualified specialists—the future quantum workforce [4], e.g., quantum engineers, since an increase in the "engineering talent flow [...] could vastly accelerate the development of quantum technologies" [5] (p. 3). According to [5] the term "quantum engineering" refers to the "application of engineering methods and principles to quantum information systems and problems" [5] (p. 3). Hence, qualification opportunities need to be implemented for this target group, both in terms of training at universities and in terms of training for professionals.

1.1. Educating the Future Quantum Workforce

Many initiatives have been implemented that aim at boosting the development of quantum ecosystems, and addressing the industry's request for a quantum-literate workforce. A topical overview of the quantum workforce landscape is provided in Ref. [6]. For example, in the US, the *Quantum Information Science and Technology Workforce Development National Strategic Plan* [7] was set up, and with the *National Q-12 Education Partnership*, a consortium has been established with the aim of preparing "America's next generation workforce with the tools to succeed in the industries of the future" [8]; see [9] for a comprehensive overview of Quantum Information Science courses at US institutions. In Europe, the *Quantum Flagship* [10] is flanked by the *Coordination and Support Action for Quantum Technology Education* (QTEdu CSA) [11], which aims at "bringing together quantum professionals from different walks of quantum education, research, and industry to help define the future of Europe's quantum workforce as the second quantum revolution unfolds" [11]. With Canada's *National Quantum Strategy* [12], the *Sydney Quantum Academy* in Australia [13], India's *National Mission on Quantum Technologies and Applications* [14], or Singapore's *Quantum Engineering Program* [15], similar efforts aimed at educating the future quantum workforce have been launched in other parts of the world.

In prior research, the industry needs for the future quantum workforce have been explored in different studies [16–18]. The results of such studies are essential for future efforts, since they may serve as a starting point for the development of target-group-specific educational programmes on quantum technologies, as has recently been shown with the development of a quantum engineering (undergraduate) programme presented in Ref. [5]. In the European context, a comprehensive Delphi study has been conducted within the QTEdu CSA, in order to collect quantum-specific competences that experts from academia and industry demand for the future quantum workforce [19,20]. The Delphi study results contributed to the development of the *European Competence Framework for Quantum Technologies* [21]. One prominent competence domain highlighted by the Competence Framework refers to quantum optics as a basis for quantum technologies, and closely related, its application in quantum photonics [21].

Quantum photonics and the underlying quantum optics fundamentals are highly relevant for almost all applications of quantum technologies, either as a core component of the application, e.g., in quantum imaging or quantum communication, or as an enabling technology [22]. Thus, quantum photonics are not limited to a specific application field. On the other hand, quantum photonics are the technology on the basis of which commercial applications can already be expected in the near future [2,22], e.g., in quantum sensor technology. For this reason, we believe that strengthening education in the field of quantum optics in (undergraduate) engineering programmes might have an impact on boosting developments in the field.

Different universities world-wide have already launched qualification programmes for engineering studies focusing on quantum optics and photonics [23–26]. Common are programmes that introduce students to the essential content domains of quantum physics via quantum optics fundamentals, most often starting from experiments with heralded photons (see [27]) for two main reasons:

1. The experimental and physical foundations of experiments with heralded photons may directly be leveraged to quantum technology applications within teaching scenarios, e.g., with regard to quantum computing [28], quantum metrology [29], or quantum information [30,31]. This content-specific argument is especially important with regard to undergraduate courses within study programmes for prospective quantum engineers.
2. Such experiments "provide the simplest method to date for demonstrating the essential mystery of quantum physics" [32] (p. 1), and "elegantly illustrate the fun-

damental concepts of quantum mechanics such as the wave-particle duality of a single photon, single-photon interference, and the probabilistic nature of quantum measurement" [33] (p. 1). Hence, such quantum optics-based approaches are likely to be conducive to circumvent widespread learning difficulties regarding quantum concepts, as has already been indicated by empirical results presented in Ref. [34].

1.2. Aim of This Study

Despite the above-mentioned advantages that potentially emerge from the integration of quantum optics courses with a focus on experiments with heralded photons into quantum engineering study programmes, no empirical results regarding the evaluation of such study programmes have been published in the literature so far. Not least, this is due to a lack of target-group- and content-domain-specific concept inventories that allow for the investigation of students' learning progression. Therefore, the development of a concept inventory that can be used to inform instructors about engineering students' mastery of quantum optics concepts constitutes a research desideratum to date. This is where this project comes in: We present a 16-item concept inventory to assess students' conceptual understanding of quantum optics, with a focus on experiments with heralded photons on a qualitative level. We administered this Quantum Optics Concept Inventory as a post-test to $N = 216$ students in the end of a four-week programme on quantum optics that is embedded into an undergraduate engineering course leading students from the basics of heralded photons experiments to the first insights into quantum technologies applications. Consequently, the objectives of the study presented in this paper are twofold: From a Rasch scaling approach, we

(1) investigate as to how our concept inventory psychometrically functions on a sample of engineering students,
(2) establish a difficulty scale regarding concepts covered by quantum optics experiments with heralded photons suggested by students' scores on the concept inventory.

In Section 2, we provide insights into the research background regarding learning about quantum optics content domains, before the development of the Quantum Optics Concept Inventory for engineering students is sketched in Section 3. The research questions are formulated in Section 4, and, in Section 5, we outline the methodology employed in our study. The results obtained with regard to the research questions are given in Section 6, and finally, we discuss our findings against the backdrop of the literature in Section 7.

2. Research Background

In recent years, there has been an increase in the number of research projects aimed at introducing students of different levels and target groups to quantum physics, starting from a quantum optical perspective [32,35–39]. Although empirical research on teaching and learning quantum physics is gaining pace in general [40], there have been no empirical studies examining engineering students' learning in such quantum optics courses to date. Contributing to closing this gap is a major goal of this study. In the next two Subsections, we present some results of research on the teaching and learning of quantum optics in secondary and undergraduate physics courses, in order to situate the study presented in this paper against the backdrop of the existing literature.

2.1. Assessment of Students' Conceptual Understanding of Quantum Optics

Various concept inventories that allow for the assessment of students' conceptual understanding of quantum physics have been published so far; a detailed overview of these concept tests aimed at the secondary school level or at the undergraduate physics level is given in Ref. [41]. These instruments either cover basic content domains of quantum physics (see [42–45]) or address aspects of the quantum formalism (see [46–48]).

The only test instrument on the introductory aspects of quantum optics is presented in Ref. [49]. This instrument consists of single-choice items, and has been developed in an iterative process, including different rounds of piloting. In terms of content, it covers both theoretical and experimental aspects of experiments with heralded photons. The content validity of the instrument has been ensured by means of an expert survey, and the tests' psycometric properties have been analyzed in terms of classical test theory, based on data from $N = 86$ undergraduate engineering students after instruction in introductory quantum optics. The test was found to enable a reliable measure ($\alpha = 0.78$), and has good item statistics. This instrument serves as a starting point for the project presented in this paper: In Section 3, we describe how we adapted it for use on a sample of engineering students.

2.2. Students' Conceptions on Quantum Optics Aspects

Two reviews provide a comprehensive overview of learning difficulties that students may encounter in quantum physics: one review focusing on secondary and lower elementary levels [41], and another one focusing on upper-level undergraduate students [50]. Currently, only a few studies with relatively small samples have been published exploring obstacles that students encounter with regards to quantum optics [33,51–54]. An exploratory interview study identifying $N = 25$ students' difficulties regarding different quantum optics content domains is reported in Ref. [55]. In [55], it was shown that (a) students are predominantly unprejudiced towards the experimental and technical foundations of quantum optics experiments, e.g., with regard to the preparation of single-photon states in experiments with heralded photons via coincident events on binary detectors, and that (b) most of the students can cope with the underlying principles [55]. These indications are also supported in further studies, e.g., exploiting the technique of probing acceptance [51]. In contrast, students specifically showed learning difficulties with regard to quantum effects and photons, which are in accordance with the findings from earlier studies into students' difficulties with aspects of quantum physics (see, e.g., [56,57]), for example:

- a part of the learners interpret single-photon interference via photons that divide and then overlap with themselves [55] (p. 216);
- a part of the respondents equate photons with waves and conclude from this the necessity of observing interference phenomena [55] (p. 221);
- some learners claim that the photons' localization would fail due to the small size of the photons [55] (p. 227);
- furthermore, a part of the participants believe that quantum anticorrelation at a beam splitter cube would be caused by photons behaving like haptical particles, being either reflected or transmitted [55] (p. 232).

With respect to the technical basics of quantum optics, learning difficulties regarding the detection of photons have been reported in the literature [51,58]: for example, a widespread misconception is that a pulse or click of a single-photon detector proves the detection of a single photon, but "this is not in agreement with the theory. A photo multiplier tube (PMT) or avalanche photo diode (APD) sends clicks, because engineers designed them to do so" [51] (p. 2).

For the development of the Quantum Optics Concept Inventory presented in this paper (see Section 3), we particularly referred to difficulties that students encounter with aspects of quantum optics that are documented in the literature, as presented in this subsection.

3. The Quantum Optics Concept Inventory for Engineering Students

We used the instrument described in Ref. [49], which is the result of a comprehensive process of development and piloting (described in Section 2), as a starting point for the development of our Quantum Optics Concept Inventory, focusing on qualitative aspects of experiments with heralded photons. We adopted the instrument for use in the primary target group of engineering students. Namely, we took two steps for the development:

(1) we updated the existing items (and the corresponding distractors) from Ref. [49], in terms of language, and
(2) beyond the content domains covered in the test of Ref. [49], namely (a) the theoretical and (b) the experimental basics of quantum optics, we developed new items to

incorporate the thematic area of the technical basics of quantum optics experiments into the instrument, since we believe this content domain to be relevant to future quantum engineers.

In total, the Quantum Optics Concept Inventory consists of 16 single-choice items. In tier one, the participants are asked to select exactly one out of three answer options. In tier two, the participants are asked to rate their answer confidence on a five-point rating scale (1 stays for "I guessed", 2 for "not sure", 3 for "undecided", 4 for "sure", and 5 for "very sure"). A point is assigned if and only if the correct answer option was chosen in tier one, plus the respondent was at least sure with the response, meaning that a minimum of 4 was chosen on the confidence scale in tier two. Comparable confidence scales have previously been used for concept inventories, and they lead to the combination of two fundamental advantages [49,59–61]: On the one hand, the confidence scale

(1) helps to minimize the effect of guessing, and hence, is "useful for gauging the quality of students' understanding" [59] (p. 3); on the other hand,
(2) it allows for the exploration of learning difficulties regarding the content area under investigation, namely by analyzing incorrect answers that were given confidently [62]. This point is postponed for the future research.

From a content-specific perspective, the Quantum Optics Concept Inventory aimed at engineering students is developed to test three content domains regarding quantum optics with heralded photons: technical-experimental foundations (domain 1), preparation of single-photon states (domain 2), and quantum effects (domain 3). Each domain is addressed by at least five items. An overview of the three content areas and the topics covered are given in Table 1.

Table 1. Overview of the content domains covered in the Quantum Optics Concept Inventory, and the topics addressed with the related items. For a didactically prepared overview of the subject-specific topics represented in the test, e.g., the anticorrelation of single-photon states (see [18,51]). The list of items is given in Appendix A.

Content Domain	No. of Items	Topics Covered (Related Items)
Technical-experimental foundations	5	Set-up/adjustment of quantum optical experiment (items 1, 2, and 4), Single-photon detection (items 5 and 8)
Preparation of single-photon states	6	Parametric downconversion and energy conservation (items 3, 6, and 7), Coincidence technique (items 9, 10, and 15)
Quantum effects	5	Interference of single quanta (items 11, 12, and 13), Anticorrelation of single-photon states (items 14 and 16)

The content validity of the Quantum Optics Concept Inventory addressing the context of experiments with heralded photons was checked by means of an expert survey with $N = 8$ faculty members [63]. For the investigation of further aspects of validity, in particular with regards to construct validity, we checked the fit of the data gathered in our study to a one-dimensional Rasch model. Therefore, we refer the reader to our methods and results in Sections 5 and 6. Beyond that, we also describe reliablity issues in the results in Section 6.

4. Research Questions

The psychometric characterization of the Quantum Optics Concept Inventory (see Section 3) with respect to the primary target group of engineering students is the main goal of this study. Hence, the first research question (RQ1) to be addressed is:

RQ1: How does the Quantum Optics Concept Inventory function on a sample of engineering students?

In the literature, qualitative approaches have been taken so far to explore difficulties that students encounter with quantum optics concepts (see Section 2). We have outlined findings from prior studies according to which students seem to be unprejudiced towards experiments with heralded photons, in particular, regarding the preparation of single-photon states, and most of the students can cope with the underlying principles. Conversely, quantum physical effects pose more difficulties to learners. Hence, to draw on earlier studies, and to shed new light on the difficulties that engineering students encounter with certain content aspects of quantum optics experiments with heralded photons using our concept inventory, we approach a clarification of the following second research question (RQ2) in this paper:

RQ2: What difficulty scale of quantum optics content aspects is suggested by students' scores on the concept inventory?

5. Methods

5.1. Test Administration and Sample

The Quantum Optics Concept Inventory used was administered to $N = 216$ second-year engineering students (144 males, 68 females, 4 not specified) as a post-test after the students had participated in a four-week program about quantum optics with heralded photons. This four-week program was integrated in a regular introductory physics lecture for future engineers. In Section 5.2, a brief overview of the four-week intervention is given while in Appendix B, we provide an in-depth description of the educational pathway and the content aspects covered. None of the participants had any prior instruction in quantum physics beyond school physics.

The participation in our study was voluntary, and was not financially recompensed. In addition, the students were informed about anonymity, as well as about processing of their data, and were asked to give their consent to participate.

5.2. Intervention

The educational pathway of the four-week program was based on an introductory quantum optics course originally developed for the secondary school level as well as for the lower undergraduate levels, and is described in detail in [51]. For the purpose of this study, this concept was adapted in terms of content and depth—specifically, two important modifications dedicated to future engineers have been made to the original version [51] for the four-week program. Namely,

(1) while the original teaching concept exclusively covered aspects of experiments with heralded photons qualitatively to make quantum effects accessible [51], during the course implemented in this study, the same topics were also covered on a formal level, as proposed in [64];
(2) to conclude, the course implemented in this study provided future engineers with an outlook on quantum technologies, namely by dealing with quantum cryptography, e.g., the BB84 protocol with single-photons [65,66].

The four-week program was divided into two consecutive parts, and culminated in the discussion of the 1986 paper by Grangier et al. [67]. Hence, from a theoretical point of view, the program was based on the theory of optical coherence, which "investigates the properties of light in terms of correlation functions between two distinct fields or of a single field at different points in time and space" [68] (p. 144).

It is noteworthy that the development of a well-grounded proposal for educating the future quantum workforce is beyond the scope of this paper. However, to provide a deeper background of the intervention implemented in this study, we sketch the essential topics that were covered in the four-week quantum optics course, in Appendix B. We reflect on the role of the educational path for the results of this study in the limitation (see Section 8.1).

5.3. Data Analysis

In the following, we present the analysis tools incorporated into our research, bundled for each research question.

5.3.1. Analysis Carried Out to Answer RQ1

In a first step, we provide a characterization of the instrument in the sense of classical test theory; i.e., Cronbach's α, as well as the items' point-biserial coefficents and items difficulties are presented. Here, we refer to the accepted ranges of 0.2 to 0.8 for the item difficulty [69] and values $\geq$0.20 for the point-biserials [70]. For Cronbach's α, values above 0.7 are considered as acceptable [71].

In a further step, the goal is to divorce statements regarding the concept inventory from the sample it was piloted with. This is enabled via dichotomous Rasch scaling. To justify this analysis, we investigated the pre-conditions of Rasch scaling (see [72]) by verifying that:

- the skewness and kurtosis of the items do not exceed the range of -2 to $+2$,
- the items are locally independent, and
- the uni-dimensionality of the construct can be assumed.

It is noteworthy that a verification of our instrument's uni-dimensionality also supports construct validity [60].

After the pre-conditions of Rasch Scaling are met, the participants' ability levels and the items' difficulties are estimated. The goodness-of-fit parameters, Outfit mean-square (MNSQ) and Infit MNSQ, further describe how well the items fit the model—their expected value is 1; values below this indicate unmodeled noise, and values above this indicate redundancy, meaning that the respective item does not contribute much to estimating the participants' latent trait levels [73]. The accepted ranges for these fit statistics are between 0.7 and 1.3 [74]. All parameters were calculated using the software R (Version 4.1.2) and its packages ltm (1.0-2) [75], TAM 4.0-16 [76], and eRm (1.0-2) [77].

5.3.2. Analysis Carried Out to Answer RQ2

In our concept test, three quantum optics content domains are addressed with at least five items each (see Table 1). To compare the difficulties of the items related to each of these content domains, we calculated their average values, μ (in logits), and their uncertainties in terms of standard errors (SEs), as has been performed in prior studies [78].

Since the average difficulty of all items is set to zero in the Rasch modeling approach, item groups with negative average item difficulties are easier to solve for the sample students than the test average, whereas item groups with positive average item difficulties are more difficult than the test average.

6. Results

6.1. Psychometric Characterization

First, a distractor analysis was conducted, which revealed that for most of the items of our concept inventory, each answer option was selected by at least 5% of the participants, which constitutes a widely used criterion in test development (see [49,60]). Next, we analyzed item difficulties and point-biserial coefficents in the sense of classical test theory (see Table 2). It can be observed that Cronbach's α increases only slightly when item 10 is omitted. The item difficulties and the point biserials show a reasonable range, overall.

Table 2. Each item's difficulty and point-biserial-coefficent, as well as the adjusted Cronbach's $\bar{\alpha}$ if the respective item is dropped. See text for details.

Item No.	Difficulty	Point-Biserial	$\bar{\alpha}$
1	0.15	0.35	0.73
2	0.37	0.30	0.73
3	0.24	0.46	0.72
4	0.91	0.22	0.74
5	0.43	0.48	0.72
6	0.63	0.39	0.73
7	0.82	0.33	0.73
8	0.37	0.29	0.74
9	0.80	0.21	0.74
10	0.59	0.20	0.75
11	0.54	0.34	0.73
12	0.20	0.34	0.73
13	0.35	0.44	0.72
14	0.32	0.31	0.73
15	0.35	0.34	0.73
16	0.61	0.36	0.73

A dichotomous Rasch model was justified by the data. Values of skewness and kurtosis are given in Table 3. Item 4 does not meet the criterion mentioned in Section 5; however, it was retained for content reasons, as we elaborate on in Section 7. Local independence was verified by checking the Q_3 correlation matrix for values above 0.2 [79,80]. Unidimensionality was confirmed using the R-package sirt (version 3.9-4), finding weighted indices DETECT $= -0.041$ (<0.20), ASSI $= -0.017$ (<0.25), and RATIO $= -0.036$ (<0.36) [81]. In summary, the pre-conditions of Rasch scaling were met.

Table 3. Overview of the relevant parameters for a dichotomous Rasch Model; "SE" stands for the standard error of item difficulty. See text for details.

Item No.	Skewness	Kurtosis	Difficulty	SE	Infit	Outfit
1	1.99	2.00	2.096	0.206	0.929	0.916
2	0.54	−1.72	0.648	0.156	1.056	1.042
3	1.22	−0.51	1.393	0.174	0.902	0.772
4	−2.83	6.07	−2.697	0.247	0.985	0.873
5	0.30	−1.93	0.364	0.152	0.902	0.889
6	−0.54	−1.72	−0.652	0.155	0.951	1.139
7	−1.63	0.67	−1.787	0.189	0.956	0.952
8	0.54	−1.72	0.648	0.156	1.079	1.048
9	−1.48	0.20	−1.649	0.183	1.080	1.109
10	−0.38	−1.87	−0.462	0.153	1.148	1.197
11	−0.15	−2.00	−0.184	0.151	1.022	1.002
12	1.48	0.20	1.647	0.183	0.987	0.907
13	0.63	−1.62	0.745	0.157	0.924	0.931
14	0.80	−1.37	0.948	0.161	1.028	1.003
15	0.63	−1.62	0.745	0.157	1.017	0.973
16	−0.46	−1.81	−0.556	0.154	1.005	0.967

The "expected a posteriori/plausible value" (EAP/PV) reliability was found to be 0.746, and the "weighted likelihood estimates" (WLE) reliability was 0.724, exceeding the lower threshold of 0.5 according to [74]. An overview of all the estimated parameters by the Rasch model is presented in Table 3. The values of the Infit MNSQ and Outfit MNSQ all lie in the accepted range indicating a good Rasch homogeneity. The Infit statistic varies between 0.902 (item 3) and 1.148 (item 10), with a mean value of $\langle \text{Infit} \rangle = 1.004$. The Outfit statistic varies between 0.772 (item 3) and 1.197 (item 10), with a mean value of

$\langle \text{Outfit} \rangle = 0.983$. A graphical representation of the findings is given in Figure 1.

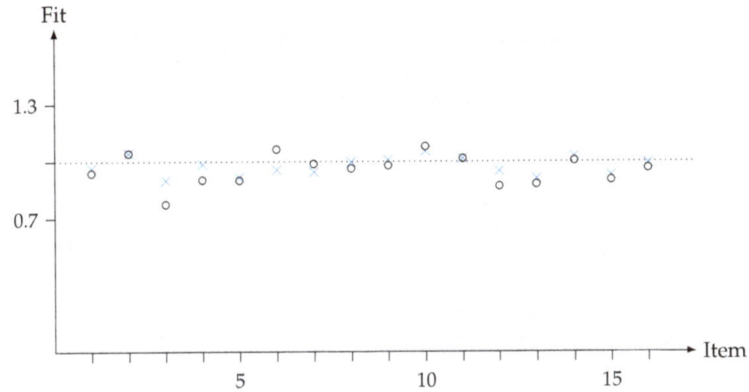

Figure 1. Infit MNSQ (crosses) and Outfit MNSQ (circles) for all items of the concept inventory. See text for details.

The item difficulties may be investigated by visualizing their relationship to the solving probability. In this regard, each item can be described by a logit function, where the solving probability is plotted against the latent trait level on a continuous scale. The resulting graph of this function is characteristic for the item, and is thus called the item characteristic curve [74]. The curves are shown in Figure 2.

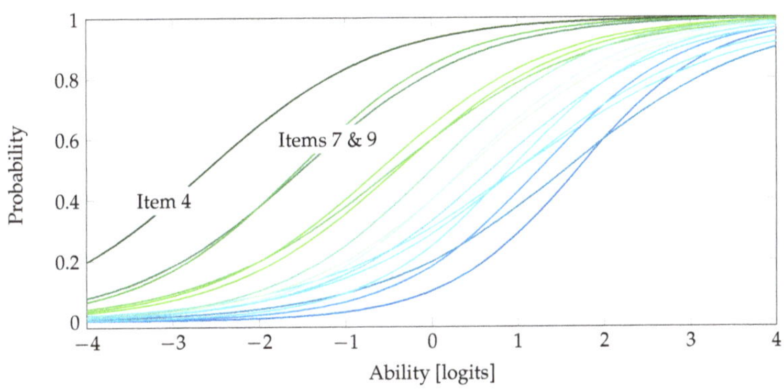

Figure 2. Item characteristic curves (ICC) for all items of the concept inventory. While the blue, light-blue, and light-green curves correspond to items that show good fit to the Rasch model, the outliers, namely items 4, 7, and 9, can also be identified. This observation is further substantiated analyzing the Wright Map presented in Figure 3.

The item difficulty ranges from −2.697 logits (item 4) to 2.096 logits (item 1). More important, however, is how the items correspond to the latent trait levels within the sample. This connection is investigated in Figure 3 using a Wright Map. The Wright Map shows the distribution of item difficulty and respondent latent trait level on a common scale, and thus enables an assessment of how well the instrument is suited to measuring different areas of ability. For example, Figure 3 shows that in densely populated areas, the item density reacts accordingly, but in the upper end of the ability scale ($\geq$2 logits), no item corresponds, meaning that our concept inventory is not suited to distinguishing between participants with very high trait levels. In other words, in the future, this concept inventory may be refined by developing at least one more difficult item.

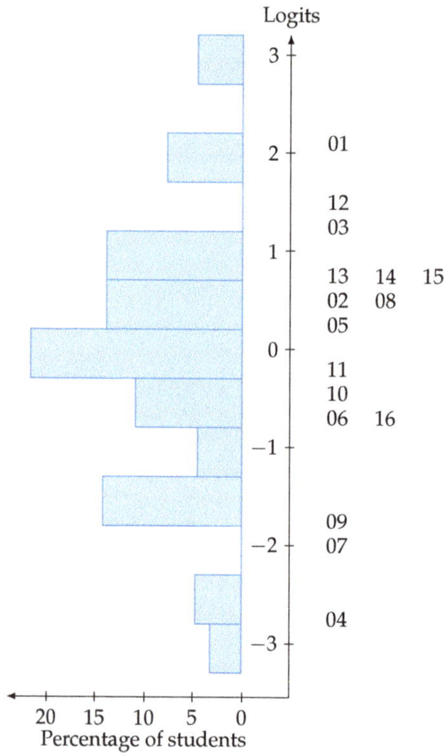

Figure 3. Wright Map of our concept inventory. The left-hand side represent the frequency of respondents' latent trait levels within the sample. The right-hand side represents a hierarchical order of all items along the logit scale.

In summary, the results presented in this Subsection may contribute to a clarification of the research question RQ1. We judge the overall psychometric quality of the Quantum Optics Concept Inventory and its functioning on our sample of engineering students in Section 7 below referring to the literature on concept inventories.

6.2. Difficulty Scale of Quantum Optics Concepts

The Rasch scaling of our concept inventory was further leveraged to investigate the mean difficulty of the different content domains covered in our diagnostic tool (see Table 1), as has been performed in prior research using concept inventories (see [82]).

The most difficult content domain appears to be domain 3 (quantum effects), with an average item difficulty of ($\mu = 0.52$, SE $= 0.35$), as is graphically shown in Figure 4. The next content domain according to difficulty is the domain 1 on technical and experimental foundations of quantum optics experiments with heralded photons ($\mu = 0.21$, SE $= 0.71$). This content domain, in total, appears to be slightly more difficult than the average. In contrast, the content domain 2 on the preparation of single-photon states ($\mu = -0.40$, SE $= 0.47$) lays well below the average difficulty, and hence, the corresponding items did not pose a problem for a majority of the students. In summary, these results allow for the establishment of a difficulty scale of quantum optics concepts with respect to experiments with heralded photons, and hence, they contribute to a clarification of research question RQ2.

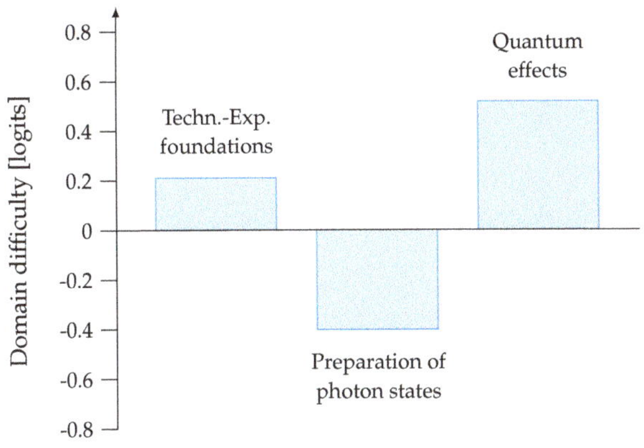

Figure 4. Average difficulties of the quantum optics concepts, evaluated by our concept inventory, measured in logits.

In Section 7 just below, we discuss the observed differences between the three content domains covered in our concept inventory with regard to their average item difficulties. There, we provide evidence that the findings of our study are consistent with prior research, and derive implications for the future. Based on the results obtained, it is not possible to resolve statistically significant differences regarding the mean item difficulties on a 5% level. This problem can be addressed by adding further items to the concept inventory covering the three content domains, respectively. This point is addressed in Section 8.1.

7. Discussion

7.1. Discussion of RQ1

From the standpoint of classical test theory, the item difficulties and point-biserials lie in the desired range for all items (except for item 4) and constitute a scale with good internal consistency expressed by Cronbach's $\alpha = 0.74$.

The Rasch analysis of the concept inventory showed a good degree of functioning for almost all test items. Only item 4 may be revisited due to its skewness and kurtosis values, even though it shows good fit statistics. The Wright Map (see Figure 3) illustrates how well the items can be ordered along the logit scale; they measure a conceptual understanding of quantum optics on different trait levels, and most items measure around the 0 logit area, which is most densely populated. We conclude that the test is well designed for the population that it is intended to be applied on. Merely for very high trait levels, at least one further item should be developed in the future. Additionally, one could make a case for also developing items in the region of about −2.5 logits and about −1.5 logits, as visual gaps can be observed both in the Wright Map, as well as in the overview of the item characteristic curves (see Figure 2). Further refinement to close this gap, and thus, to contribute to an overall continuous item difficulty distribution may result in a further refined concept inventory for quantum optics in engineering curricula.

In essence, only item 4 turned out to be problematic. It seemed too easy for our sample, from the perspective of classical test theory (item difficulty 0.91). In addition, its high skewness and kurtosis suggest a deviation from the normal distribution. However, the Wright Map substantiates that items of such difficulty (−2.697 logits from the Rasch perspective) are in fact necessary, and no other item measures similarly at the lower end of the scale. Furthermore, item 4 is important for content reasons, since it is the only item of the instrument that asks for a differentiation between single-photon from further types of quanta, e.g., single electrons, in this case. The mix-up of photons and electrons, however, has been proven to be a learning obstacle to students in [83,84]. To this end, this item is to be retained in order to ensure content validity. Additionally, such outlier items are considered in a categorial judgement scheme provided in [70]. We use this judgement scheme to summarize the above discussed observations, allowing for a holistic judgement regarding all relevant concept inventory characteristics. In this regard, we follow Ref. [60], where this judgement scheme has been expanded to Rasch scaling parameters. The overall quality of our concept inventory is summarized in Table 4.

Table 4. Categorial judgment scheme and assignment rules for evaluating a concept inventory according to [70]. Values in parenthesis indicate the number of items that can fall outside of this recommendation. The judgement of our concept inventory is presented in the last column labeled QOCI.

Analysis	Excellent	Good	Average	Poor	QOCI
Classical Test theory					
Item Statistics					
Difficulty	0.2–0.8	0.2–0.8 (3)	0.1–0.9	0.1–0.9 (3)	good
Discrimination	>0.2	>0.1	>0	>−0.2	excellent
Total score reliability					
α of total score	>0.9	>0.8	>0.65	>0.5	average
α-with-item-deleted	All items less than overall α	(3)	(6)	(9)	good
Item Response Theory					
Individual item measures					
Infit MNSQ	0.7–1.3	0.6–1.4	0.5–1.5	–	excellent
Outfit MNSQ	0.7–1.3	0.6–1.4	0.5–1.5	–	excellent
All items fit the model	(2)	(4)	(6)	(8)	excellent

7.2. Discussion of RQ2

The Quantum Optics Concept Inventory is developed to test students' conceptual understanding of quantum optics basics, with a focus on experiments with heralded photons focusing on three content domains, namely technical-experimental foundations (domain 1), the preparation of single-photon states (domain 2), and quantum effects (domain 3).

The data analysis presented in Section 6 revealed that the content domain 3 addressing quantum effects was the most challenging one for the study participants (average item difficulty, $\mu = 0.52$). The content domain was covered by six items, three of which address single-photon interference (items 11, 12, and 13), and two of which focus on single-photons and the anticorrelation effect (items 14 and 16). Analyzing these items in more detail, two observations become striking.

(1) The item difficulties of the items 12 (1.647 logits), and 13 (0.745 logits) both lay well above average—hence, single-photon interference obviously poses conceptual difficulties to the study participants. This observation is in accordance with prior research, and can be enriched by findings from various qualitative studies exploring student difficulties on the interference of single quanta [33,55].

(2) Item 14, focusing on the anticorrelation effect of single-photon states at an optical beam splitter cube, has an item difficulty of 0.948 logits. Hence, the concept that photons are either transmitted or reflected at a beam splitter cube seems to be difficult for students. This observation fits well with results presented in an earlier contribution [51], where the authors used a micro-intervention to explore students' understanding of single-photons' anticorrelation, using the technique of probing acceptance. Beyond that, this observation is particularly striking against the background of the finding described above regarding single-photon interference: while items 12 and 13 address the wave nature of photons, item 14 can be associated with a particle nature of photons. Hence, the three items represent one of the fundamental issues of quantum physics: wave-particle duality.

Taken together, the observations fit well with previous studies into students' conceptual hurdles regarding the wave-particle duality of quantum entities (see [55,85,86]; for a review of studies on this topic, see [41]). In summary, it is not a surprise that in Ref. [87], the authors demand "a radical reconstruction" (p. 257) of learners' initial knowledge when it comes to learning about quantum physics.

The items related to the content domain 1 on technical-experimental foundations had an average item difficulty of $\mu = 0.21$. The content domain is covered by five items of the concept inventory, which cover the whole logit scale regarding item difficulties, ranging from −2.697 logits (item 4) at the lower end of the difficulty scale, to 2.096 logits (item 1) at the upper end. While the more difficult items of this content domain (item 1, with 2.096 logits, and item 2, with 0.648 logits) cover experimental aspects, e.g., the use of beam splitter cubes (item 1) or the adjustment of optical experiments (item 2), the other items covering technical aspects appear to have low-to-average item difficulties (−2.697 logits for item 4, 0.364 logits for item 5, and 0.648 logits for item 8).

Finally, the items related to the content domain 2 on the preparation of single-photon states had an average item difficulty of $\mu = -0.40$, and hence, were easier than average. This observation is in accordance to the results of prior (qualitative) studies, which revealed that students can predominantly cope well with the basic ideas underlying the preparation of single-photon states in experiments with heralded photons via coincident events on binary detectors [51,55]. To this end, we believe that the results obtained provide a further argument in favor of using the experimental setting of heralded photons as a gateway towards quantum effects in teaching quantum physics. The easiest items related to this content domain were items 6 (−0.652 logits) and 7 (−1.787 logits), addressing the concept of energy levels and energy conservation in the context of nonlinear optics. In contrast, the most difficult, item 3 (1.393 logits), referred to parametric downconversion (PDC), a nonlinear effect leading to the emission of photon pairs from a nonlinear crystal irradiated by a laser at a certain wavelength. An in-depth analysis of the students' response behavior on the answer options of item 3 shows that the students tend to conceptually mess up PDC with reflection and diffraction from geometrical optics.

8. Conclusions

We presented a 16-item Quantum Optics Concept Inventory aimed at assessing engineering students' conceptual understandings of the fundamental aspects of quantum optics experiments with heralded photons. This tool has been adopted from an instrument previously published elsewhere [49]. We added items related to the technical aspects of single-photon experiments, e.g., addressing single-photon detection, making it a valuable tool for evaluating teaching-learning scenarios

- aimed at engineering university students, and
- designed to introduce these students to quantum physics and modern quantum technologies via quantum optics experiments with heralded photons.

8.1. Limitations

There are some limitations of our study that may serve as a springboard for future research. These limitations either stem from the study design itself (Section 8.1.1), or are related to the concept inventory evaluated in this study (Section 8.1.2).

8.1.1. Limitations of this Study

Without a doubt, the evaluation of the Quantum Optics Concept Inventory presented in this paper is—at least to some extent—dependent on the quantum optics instruction that the study participants had received prior to test administration. This a known concern in educational research in general [88]—especially when it comes to piloting novel assessment tools. However, in the current study, this issue was mitigated by means of Rasch scaling utilized for the psychometric characterization of the Quantum Optics Concept Inventory, since the Rasch model permits the separation of person and item parameters [89].

Against the backdrop of the above argumentation, it becomes apparent that the difficulty scale of quantum optics concepts established on the basis of the participants' test results (see Section 6.2) should not be confused with statements about the perceived difficulty of quantum optics topics. Rather, this difficulty scale should be understood to shed more light on (a) the items of the Quantum Optics Concept Inventory, and (b) on possible learning difficulties that might pose obstacles to student learning in the context of introductory quantum optics (see Section 7.2). Taken together, the results presented in this study may not only lead to a revision of the Quantum Optics Concept Inventory, but they might also initiate the refinement of the four-week program implemented in a lecture for future engineers in the course of this study.

8.1.2. Limitations Related to the Quantum Optics Concept Inventory

We merged the viewpoints of classical test theory with those of probabilistic test theory to provide a psychometric characterization of the Quantum Optics Concept Inventory. Based on the data gathered from $N = 216$ engineering students who took the Quantum Optics Concept Inventory as a post-test after instruction, we found our instrument to exhibit average-to-excellent psychometric properties following the categorical judgment scheme for evaluating a concept inventory of Ref. [70]. In particular, the Rasch analysis results revealed that the items discriminate well among students from lower to medium–high competence levels (see Figure 3). However, the items do not manage to discriminate well between students at the upper end of the ability scale. Hence, in order to refine our instrument accordingly, items with a higher difficulty should be added.

Analyzing the data in more detail with respect to the content domains covered in our inventory (see Table 1), findings from prior research could be replicated, adding further evidence for the functioning of our concept inventory: The average item difficulty for the items related to the content domain 3 on quantum effects lies well above the average. Specifically, we found that items addressing different aspects of wave-particle duality, i.e., items 12, 13, and 14, covering single-photon interference and the anticorrelation of single-photon states, were found at the upper end of the difficulty scale. This is consistent with the wave-particle duality posing several conceptual hurdles to learners from different target groups, as is well documented in the literature [41]. However, more differentiated statements on student conceptions, especially regarding quantum optics concepts, are difficult to make with the current version of our concept inventory: since for now, all topics are covered with only two or three items each, further items would be required in order to allow for a more comprehensive insight into students' conceptions regarding the respective isolated topics. In contrast, the brevity of our concept inventory allows for economic practical use aimed at the evaluation of teaching–learning environments. Against this backdrop, the creation of additional items is always to be considered carefully, in order to maintain the practical applicability of the instrument.

Finally, at the qualitative level, an important limitation must be considered, which stems from the fact that the items of the Quantum Optics Concept Inventory represent a compromise of the following criteria:

(i) intelligible wording,
(ii) coverage of overarching concepts of quantum optics in the heralded photon realm, rather than addressing all the details of a topic,
(iii) subject-specific correctness.

Item 7 of the Quantum Optics Concept Inventory (see Appendix A), which deals with PDC, may serve as a prime example of this tradeoff: While PDC photons in general have different frequencies, in this item, we focus on the particular situation where the two PDC photons have equal frequencies. Hence, this item is intended to evaluate whether students have understood that energy is conserved in the PDC process. In contrast, assessing students' conceptual understanding of details of phasematching conditions is outside the scope of this item (see bullet 2 in the above list). Again, developing additional items that cover further topics and aspects related to quantum optics experiments with heralded photons could help alleviate the constraints imposed by the aforementioned compromise.

8.2. Outlook

In the future, standardized feedback formats need to be established, such that outcomes can easily be communicated to various stakeholders, e.g., to teachers or to their students [90]. For this purpose, instead of presenting statistics based on raw test values, feedback formats that map the distribution of student performance on categorically graded levels of proficiency are required. Future studies may draw on the determination of such levels of proficiency with respect to the content area under investigation, in order to describe (and predict) students' learning progressions [91].

The number of proficiency levels regarding quantum optics experiments with heralded photons that can be empirically separated using our concept inventory depends on the person separation reliability of our concept inventory. According to [92], the person seperation reliability, s, can be used to calculate the G-value,

$$G = \sqrt{\frac{s}{1-s}},$$

which allows for the determination of the number of empirically separable proficiency levels via

$$\left[\frac{4G+1}{3}\right].$$

Here, $[-]$ is the closest integer; for details, see [74]. For the Quantum Optics Concept Inventory under investigation, we obtained $s = 0.75$, and hence, $G = \sqrt{3}$, meaning that it allows us to differentiate between

$$\left[\frac{4G+1}{3}\right] = 3$$

proficiency levels among engineering students. After a revision of our instrument to tackle the above-mentioned weaknesses, future research will address the definition of the proficiency levels regarding quantum optics experiments with heralded photons based on data from larger samples.

Author Contributions: Conceptualization, all authors; methodology, all authors; formal analysis, all authors; investigation, all authors; writing—original draft preparation, all authors; writing—review and editing, all authors; visualization, all authors. All authors have read and agreed to the published version of the manuscript.

Funding: We acknowledge financial support by Deutsche Forschungsgemeinschaft and Friedrich-Alexander-Universität Erlangen-Nürnberg within the funding programme "Open Access Publication Funding".

Institutional Review Board Statement: Ethical review and approval were waived for this study, due to the fact that the study was in accordance with the Local Legislation and Institutional Requirements: Research Funding Principles (https://www.dfg.de/en/research_funding/principles_dfg_funding/research_data/index.html, (accessed on 15 September 2022)), and General Data Protection Regulation (https://www.datenschutz-grundverordnung.eu/wp-content/uploads/2016/04/CONSIL_ST_5419_2016_INIT_EN_TXT.pdf (accessed on 15 September 2022)).

Informed Consent Statement: Informed consent was obtained from all subjects involved in the study to publish this paper.

Data Availability Statement: The data presented in this study are available on request from the corresponding author.

Conflicts of Interest: The authors declare no conflicts of interest.

Appendix A. The Quantum Optics Concept Inventory

(The correct answer option is marked by an asterisk.)

Item 1. A beam splitter

(a) is employed in the Michelson interferometer, because it can be used to split an incident ray of light into two partial beams.
(b) is made out of two merged prisms, where one of them is responsible for the transmitted beam and one for the reflected beam.
*(c) separates incident rays of light or superimposes two rays of light.

Item 2. To precisely adjust a laser beam along a specific straight line on a breadboard in a quantum optics experiment,

(a) two mirrors can be used to deflect the laser on this line.
*(b) two apertures can be used to define two points laying on this line.
(c) two lenses can be used to focus the laser beam on this line.

Item 3. If a non-linear crystal is irradiated by laser light, then,

*(a) light is emitted by the crystal.
(b) the laser beam is divided in two beams.
(c) diffraction leads to laser light forming a cone after passing the crystal.

Item 4. Experiments with heralded photons differ from experiments with single electrons, because:

(a) electrons are bigger than photons, leading to technical difficulties.
(b) conducting the double-slit experiment one after the other with single electrons does not lead to an interference pattern.
*(c) experiments with single electrons require a vacuum.

Item 5. The human eye may not be used as a detector for single photons, because:

*(a) the intensity of the light falling on the eye in the single-photon regime ist too low.
(b) the eye can only perceive light at specific wavelengths, but not those of single photons.
(c) the photons are too small to be resolved with human eyes.

Item 6. If an electron transitions from an energy state E_1 to an energy state $E_2 < E_1$, light is emitted. The bigger the energy difference, $\Delta E = E_1 - E_2$, (adopted from [43])

(a) the more photons are emitted.
(b) the longer the wavelength of the emitted light.
*(c) the shorter the wavelength of the emitted light.

Item 7. When irradiated with laser light at a certain wavelength, parametric downconversion can be driven in a nonlinear crystal. This process leads to the emission of photon pairs. Both photons have

(a) half the wavelength of the incident laser light.
(b) the same wavelength of the incident light of the laser.
*(c) double the wavelength of the incident light of the laser.

Item 8. Avalanche photodiodes used for single-photon detection

(a) count the number of registered photons within a certain time interval.
*(b) lead to electron avalanches when some kind of energy portion is registered.
(c) point to the detection of a single-photon with each click.

Item 9. Coincident events measured at avalanche photodiodes in experiments with heralded photons

*(a) mark simultaneous clicks at two or more detectors.
(b) represent a measure of the detectors' dark count rate.
(c) can each be associated with the detection of a single-photon state.

Item 10. The coincidence technique is used in experiments with heralded photons

(a) to show that photons are physical entities with finite size.
(b) in order to experiment with two single-photons simultaneously.

*(c) for single-photon detections.

Item 11. When the double-slit experiment is repeatedly performed with only one single-photon in the apparatus at a time,

*(a) an interference pattern with minima and maxima can be observed.
(b) two well-defined detection zones can be observed.
(c) two well-defined detection zones and a zero-order maximum can be observed.

Item 12. In the Michelson interferometer, (answer options adopted from [34,45])

(a) the single-photon follows a specific path, regardless of whether I observe this path or not.
(b) the current position of a photon between source and detector is not indeterminate in principle, but unknown to the experimenter.
*(c) the photon behaves like a particle and like a wave. It is none of them.

Item 13. Which statement about the behavior of a single-photon state in an interferometer is correct? (answer options adopted from [34,45])

*(a) No one can say with certainty at which output port of the beam splitter cube a single-photon will be detected.
(b) A single-photon state is divided at the beam splitter cube.
(c) In the interferometer, the single-photon state is found in a superposition state of both, particle and wave.

Item 14. Anticorrelation of single-photon states can be observed at the outputs of a beam splitter cube because

(a) more coincident events are detected between the outputs of the beam splitter than can be expected at random.
*(b) a single-photon may only be detected once.
(c) a single photon can be in both states, 'reflected' and 'transmitted', at the same time.

Item 15. For the preparation of single-photon states in experiments with heralded photons, one needs

*(a) exactly two single-photon detectors.
(b) at least one single-photon detector.
(c) at least three single-photon detectors.

Item 16. Single-photons can be regarded as

(a) spherical entities moving along a wavy path with the speed of light.
(b) elementary energy portions of light surrounded by a wave that is responsible for interference.
*(c) indivisible energy portions of light that are never detected on both output ports of a beam splitter cube simultaneously.

Appendix B. Overview of the Four-Week Program on Introductory Quantum Optics

Appendix B.1. Part I: Foundations of Quantum Optics Experiments with Heralded Photons

The course's first part covered technical and experimental key ideas regarding experiments with heralded photons in order to help the students to make sense out of the experimental results in the later stage of the course:

- First, optical components widely used in quantum optics labs were introduced: For example, the role of mirrors for the adjustment of quantum optics experiments or the action of a beam splitter on incident light (see, e.g., [93,94]) were described.
- Second, the study participants were introduced to the properties of avalanche photodiodes operating above their breakdown voltage—also referred to as single-photon avalanche diode (SPAD) [95,96]: For example, quantum efficiency, dark count rate, and dead time [68] were covered. SPADs are binary detectors, which means that the "outcome of these APD's is either 'off' (no photons detected) or 'on,' i.e., a 'click,' indicating the detection of one or more photons" [97] (p. 1).

- Third, the students were introduced to spontaneous parametric downconversion (PDC) [98], a quantum electrodynamic process in which "an incoming pump photon decays, under energy and momentum conservation, into a photon-pair" [99] (p. 351). Here, the students learned that a) PDC is driven by irradiating a nonlinear crystal (e.g., β-barium borate) with a pump beam, most often emitted by a laser, and that b) the "spectral properties of PDC states are governed by the phasematching properties of the nonlinear material, and this determines the frequencies of the downconverted photons" [100] (p. 3442). For an comprehensive overview of parametric downconversion, we refer the reader to [101].

Appendix B.2. Part II: Experiments with Heralded Photons

The course's second part made use of the foundation that was laid in the first part: On the one hand, the students were introduced to the preparation of single-photon Fock states. On the other hand, they got in touch with quantum effects that indeed require the quantization of the electromagnetic field. To provide engineering students with insights into real quantum laboratories, interactive screen experiments of heralded photon experiments previously developed by Bronner et al. [37] were shown throughout the lectures.

- First, the students learned about heralded single-photon state preparation, which is widely used in quantum optics research [102–105]). That is, one of the PDC photons is detected and heralds the second photon at a spatially separated detector—hence, simultaneous clicks at two detectors, also referred to as a coincident events, are "taken as preparation and detection of a single photon state" [37] (p. 348).
- Second, heralded single-photon states incident on a 50:50 beam splitter were investigated. In this experiment, the lack of coincident events at the output ports of the beam splitter, i.e., photon antibunching [106], was discussed. This anticorrelation effect is irreconcilable with any classical description of light: "A single photon can only be detected once" [67] (p. 173). These observations were further substantiated on a formal level a) by means of the second-order correlation function $g^{(2)}$, which allows for a judgement of the purity of single-photon states [94], and b) by highlighting the ideas of quantum superposition and quantum random, as has been achieved in [31].
- The investigation and quantum description of Grangier et al.'s experiment from 1986 [67] represents the last step of the course. In this experiment, the students realized that by using only one single-photon state, the anticorrelation at a 50:50 beam splitter appeared simultaneously to the single-photon interference observed in a spatially separated interferometer (e.g., a Michelson interferometer) using the same single-photon state in the same experimental set-up. The students experienced that the "quantum interference phenomenon shown experimentally is a consequence of the interplay of superposition and nonlocality" [64] (p. 17), while the idea of the photon as a localizable particle is not valid. Instead, in this course, the photon was introduced as an elementary field mode excitation in the sense of quantum electrodynamics [83,84,107,108].
- In the outlook, the engineering students were given first insights into quantum technologies 2.0, namely by applying an understanding of heralded photons experiments to the context of quantum cryptography.

References

1. Dowling, J.P.; Milburn, G.J. Quantum technology: The second quantum revolution. *Philos. Trans. R. Soc. Lond. Ser. A Math. Phys. Eng. Sci.* **2003**, *361*, 1655–1674. [CrossRef] [PubMed]
2. Acín, A.; Bloch, I.; Buhrman, H.; Calarco, T.; Eichler, C.; Eisert, J.; Esteve, D.; Gisin, N.; Glaser, S.J.; Jelezko, F.; et al. The quantum technologies roadmap: A European community view. *New J. Phys.* **2018**, *20*, 080201. [CrossRef]
3. Foti, C.; Anttila, D.; Maniscalco, S.; Chiofalo, M. Quantum physics literacy aimed at K12 and the general public. *Universe* **2021**, *7*, 86. [CrossRef]
4. Venegas-Gomez, A. The quantum ecosystem and its future workforce: A journey through the funding, the hype, the opportunities, and the risks related to the emerging field of quantum technologies. *PhotonicsViews* **2020**, *17*, 34–38. [CrossRef]

5. Asfaw, A.; Blais, A.; Brown, K.R.; Candelaria, J.; Cantwell, C.; Lincoln, D.C.; Combes, J.; Debroy, D.M.; Donohue, J.M.; Economou, S.E.; et al. Building a quantum engineering undergraduate program. *IEEE Trans. Educ.* **2022**, *65*, 220–242. [CrossRef]
6. Kaur, M.; Venegas-Gomez, A. Defining the quantum workforce landscape: A review of global quantum education initiatives. *Opt. Eng.* **2022**, *61*, 081806. [CrossRef]
7. Stephen Binkley, J. et al. [Subcommittee on Quantum Information Science]; Collins, F. et al. [Committee on Science]; Lander, E. et al. [National Science & Technology Council]. *Quantum Information Science and Technology Workforce Development National Strategic Plan*; Technical Report; United States Government: Washington, DC, USA, 2022. Available online: https://www.quantum.gov/scqis-releases-strategic-plan-for-qist-workforce/ (accessed on 15 September 2022).
8. National Q-12 Education Partnership. Growing the Quantum Workforce. Available online: https://q12education.org/ (accessed on 15 September 2022).
9. Cervantes, B.; Passante, G.; Wilcox, B.R.; Pollock, S.J. An overview of Quantum Information Science courses at US institutions. In *Proceedings of the Physics Education Research Conference 2021: 2021 PERC, Virtual, 4–5 August 2021*; Bennett, M.B., Frank, B.W., Vieyra, R., Eds.; American Association of Physics Teachers: College Park, MD, USA, 2021, pp. 93–98. [CrossRef]
10. Quantum Flagship. The Future Is Quantum. Available online: https://qt.eu/ (accessed on 15 September 2022).
11. Quantum Technology Education. European Open Portal: Community and Resources. Available online: https://qtedu.eu/ (accessed on 15 September 2022).
12. National Quantum Strategy. Available online: https://ised-isde.canada.ca/site/national-quantum-strategy/en (accessed on 15 September 2022).
13. Sydney Quantum Academy. Available online: https://www.sydneyquantum.org/ (accessed on 15 September 2022).
14. Quantum Technologies. Available online: https://www.psa.gov.in/technology-frontiers/quantum-technologies/346 (accessed on 15 September 2022).
15. QEP: The Quantum Engineering Program. 2022. Available online: https://qepsg.org/ (accessed on 15 September 2022).
16. Hughes, C.; Finke, D.; German, D.A.; Merzbacher, C.; Vora, P.M.; Lewandowski, H.J. Assessing the needs of the quantum industry. *IEEE Trans. Educ.* **2022**. [CrossRef]
17. Aiello, C.D.; Awschalom, D.D.; Bernien, H.; Brower, T.; Brown, K.R.; Brun, T.A.; Caram, J.R.; Chitambar, E.; Felice, R.D.; Edmonds, K.M.; et al. Achieving a quantum smart workforce. *Quant. Sci. Technol.* **2021**, *6*, 030501. [CrossRef]
18. Fox, M.F.J.; Zwickl, B.M.; Lewandowski, H.J. Preparing for the quantum revolution: What is the role of higher education? *Phys. Rev. Phys. Educ. Res.* **2020**, *16*, 020131. [CrossRef]
19. Gerke, F.; Müller, R.; Bitzenbauer, P.; Ubben, M.; Weber, K.A. Requirements for future quantum workforce—A Delphi study. *J. Phys. Conf. Ser.* **2022**, *2297*, 012017. [CrossRef]
20. Greinert, F.; Müller, R.; Bitzenbauer, P.; Ubben, M.S.; Weber, K.A. The future quantum workforce: Competences, requirements and forecasts. *arXiv* **2022**, arXiv:2208.08249.
21. Greinert, F.; Müller, R. *Competence Framework for Quantum Technologies: Methodology and Version History*; European Commission: Brussels, Belgium, 2021. [CrossRef]
22. Moody, G.; Sorger, V.J.; Blumenthal, D.J.; Juodawlkis, P.W.; Loh, W.; Sorace-Agaskar, C.; Jones, A.E.; Balram, K.C.; Matthews, J.C.F.; Laing, A.; et al. 2022 Roadmap on integrated quantum photonics. *J. Phys. Photonics* **2022**, *4*, 012501. [CrossRef]
23. Leibniz Universität Hannover. Quantum Engineering (Master of Science). Available online: https://tinyurl.com/bdftkpec (accessed on 15 September 2022).
24. Rainò, G.; Novotny, L.; Frimmer, M. Quantum engineers in high demand. *Nat. Mater.* **2021**, *20*, 1449. [CrossRef]
25. Universität Stuttgart. Photonic Engineering. Master of Science. Available online: https://www.uni-stuttgart.de/en/study/study-programs/Photonic-Engineering-M.Sc-00001./ (accessed on 15 September 2022).
26. Princeton UniversityElectrical and Computer Engineering. Redesigned Quantum Optics Course Brings Applications to the Heart of the Experience. Available online: https://tinyurl.com/4uw4j7nn (accessed on 15 September 2022).
27. Duarte, F. *Quantum Optics for Engineers*; CRC Press: Boca Raton, FL, USA, 2017. [CrossRef]
28. O'Brien, J.L. Optical Quantum Computing. *Science* **2007**, *318*, 1567–1570. [CrossRef] [PubMed]
29. von Helversen, M.; Böhm, J.; Schmidt, M.; Gschrey, M.; Schulze, J.H.; Strittmatter, A.; Rodt, S.; Beyer, J.; Heindel, T.; Reitzenstein, S. Quantum metrology of solid-state single-photon sources using photon-number-resolving detectors. *New J. Phys.* **2019**, *21*, 035007. [CrossRef]
30. Förtsch, M.; Fürst, J.U.; Wittmann, C.; Strekalov, D.; Aiello, A.; Chekhova, M.V.; Silberhorn, C.; Leuchs, G.; Marquardt, C. A versatile source of single photons for quantum information processing. *Nat. Commun.* **2013**, *4*, 1818. [CrossRef] [PubMed]
31. Bronner, P.; Strunz, A.; Silberhorn, C.; Meyn, J.P. Demonstrating quantum random with single photons. *Eur. J. Phys.* **2009**, *30*, 1189–1200. [CrossRef]
32. Pearson, B.J.; Jackson, D.P. A hands-on introduction to single photons and quantum mechanics for undergraduates. *Am. J. Phys.* **2010**, *78*, 1422. [CrossRef]
33. Marshman, E.; Singh, C. Investigating and improving student understanding of quantum mechanics in the context of single photon interference. *Phys. Rev. Phys. Educ. Res.* **2017**, *13*, 010117. [CrossRef]
34. Bitzenbauer, P. Effect of an introductory quantum physics course using experiments with heralded photons on preuniversity students' conceptions about quantum physics. *Phys. Rev. Phys. Educ. Res.* **2021**, *17*, 020103. [CrossRef]

35. Galvez, E.J.; Holbrow, C.H.; Pysher, M.J.; Martin, J.W.; Courtemanche, N.; Heilig, L.; Spencer, J. Interference with correlated photons: Five quantum mechanics experiments for undergraduates. *Am. J. Phys.* **2005**, *73*, 127–140. [CrossRef]
36. Thorn, J.J.; Neel, M.S.; Donato, V.W.; Bergreen, G.S.; Davies, R.E.; Beck, M. Observing the quantum behavior of light in an undergraduate laboratory. *Am. J. Phys.* **2004**, *72*, 1210–1219. [CrossRef]
37. Bronner, P.; Strunz, A.; Silberhorn, C.; Meyn, J.P. Interactive screen experiments with single photons. *Eur. J. Phys.* **2009**, *30*, 345–353. [CrossRef]
38. Bitzenbauer, P. Practitioners' views on new teaching material for introducing quantum optics in secondary schools. *Phys. Educ.* **2021**, *56*, 055008. [CrossRef]
39. Scholz, R.; Friege, G.; Weber, K.A. Undergraduate quantum optics: Experimental steps to quantum physics. *Eur. J. Phys.* **2018**, *39*, 055301. [CrossRef]
40. Bitzenbauer, P. Quantum physics education research over the last two decades: A bibliometric analysis. *Educ. Sci.* **2021**, *11*, 699. [CrossRef]
41. Krijtenburg-Lewerissa, K.; Pol, H.; Brinkman, A.; van Joolingen, W. Insights into teaching quantum mechanics in secondary and lower undergraduate education. *Phys. Rev. Phys. Educ. Res.* **2017**, *13*, 010109. [CrossRef]
42. Wuttiprom, S.; Sharma, M.D.; Johnston, I.D.; Chitaree, R.; Soankwan, C. Development and use of a conceptual survey in introductory quantum physics. *Int. J. Sci. Educ.* **2009**, *31*, 631–654. [CrossRef]
43. McKagan, S.B.; Perkins, K.K.; Wieman, C.E. Design and validation of the Quantum Mechanics Conceptual Survey. *Phys. Rev. Phys. Educ. Res.* **2010**, *6*, 020121. [CrossRef]
44. di Uccio, U.S.; Colantonio, A.; Galano, S.; Marzoli, I.; Trani, F.; Testa, I. Design and validation of a two-tier questionnaire on basic aspects in quantum mechanics. *Phys. Rev. Phys. Educ. Res.* **2019**, *15*, 010137. [CrossRef]
45. Müller, R.; Wiesner, H. Teaching quantum mechanics on an introductory level. *Am. J. Phys.* **2002**, *70*, 200–209. [CrossRef]
46. Cataloglu, E.; Robinett, R.W. Testing the development of student conceptual and visualization understanding in quantum mechanics through the undergraduate career. *Am. J. Phys.* **2002**, *70*, 238–251. [CrossRef]
47. Goldhaber, S.; Pollock, S.; Dubson, M.; Beale, P.; Perkins, K.; Sabella, M.; Henderson, C.; Singh, C. Transforming upper-division quantum mechanics: Learning goals and assessment. *AIP Conf. Proc.* **2009**, *1179*, 145–148. [CrossRef]
48. Marshman, E.; Singh, C. Validation and administration of a conceptual survey on the formalism and postulates of quantum mechanics. *Phys. Rev. Phys. Educ. Res.* **2019**, *15*, 020128. [CrossRef]
49. Bitzenbauer, P. Development of a test instrument to investigate secondary school students' declarative knowledge of quantum optics. *Eur. J. Sci. Math. Educ.* **2021**, *9*, 57–79. [CrossRef]
50. Singh, C.; Marshman, E. Review of student difficulties in upper-level quantum mechanics. *Phys. Rev. Phys. Educ. Res.* **2015**, *11*, 020117. [CrossRef]
51. Bitzenbauer, P.; Meyn, J.P. A new teaching concept on quantum physics in secondary schools. *Phys. Educ.* **2020**, *55*, 055031. [CrossRef]
52. Bitzenbauer, P.; Meyn, J.P. Fostering students' conceptions about the quantum world—Results of an interview study. *Prog. Sci. Educ.* **2021**, *4*, 40–51. [CrossRef]
53. Ubben, M.S.; Bitzenbauer, P. Two cognitive dimensions of students' mental models in science: Fidelity of gestalt and functional fidelity. *Educ. Sci.* **2022**, *12*, 163. [CrossRef]
54. Waitzmann, M.; Scholz, R.; Weßnigk, S. Wirkung eines Realexperiments auf quantenphysikalische Argumentation. Unsicherheit als Element von naturwissenschaftsbezogenen Bildungsprozessen. In *Proceedings of the Jahrestagung der Gesellschaft für Didaktik der Chemie und Physik, Online, 13–16 September 2021*; Universität Duisburg-Essen: Duisburg, Germany, 2022; p. 42.
55. Bitzenbauer, P. *Quantenoptik an Schulen. Studie im Mixed-Methods Design zur Evaluation des Erlanger Unterrichtskonzepts zur Quantenoptik*; Logos Verlag: Berlin, Germany, 2020. [CrossRef]
56. Ireson, G. The quantum understanding of pre-university physics students. *Phys. Educ.* **2000**, *35*, 15–21. [CrossRef]
57. Ireson, G. A multivariate analysis of undergraduate physics students' conceptions of quantum phenomena. *Eur. J. Phys.* **1999**, *20*, 193–199. [CrossRef]
58. Donhauser, A.; Bitzenbauer, P.; Meyn, J.P. Von Schnee- und Elektronenlawinen: Entwicklung eines Erklärvideos zu Einzelphotonendetektoren. *PhyDid B—Didaktik der Physik—Beiträge zur DPG-Frühjahrstagung* **2020**, *Bonn*, 235–240. Available online: https://ojs.dpg-physik.de/index.php/phydid-b/article/view/1021 (accessed on 15 September 2022).
59. Aslanides, J.S.; Savage, C.M. Relativity concept inventory: Development, analysis, and results. *Phys. Rev. Phys. Educ. Res.* **2013**, *9*, 010118. [CrossRef]
60. Veith, J.M.; Bitzenbauer, P.; Girnat, B. Assessing learners' conceptual understanding of introductory group theory using the CI^2GT: Development and analysis of a concept inventory. *Educ. Sci.* **2022**, *12*, 376. [CrossRef]
61. Urlacher, M.A.; Brown, S.A.; Steif, P.S.; Bornasal, F.B. Practicing civil engineers' understanding of statics concept inventory questions. In Proceedings of the 2015 ASEE Annual Conference and Exposition, Seattle, WA, USA, 14–17 June 2015; pp. 26.1236.1–26.1236.12. [CrossRef]
62. Hasan, S.; Bagayoko, D.; Kelley, E.L. Misconceptions and the Certainty of Response Index (CRI). *Phys. Educ.* **1999**, *34*, 294–299. [CrossRef]

63. Bitzenbauer, P.; Meyn, J.P. Inhaltsvalidität eines Testinstruments zur Erfassung deklarativen Wissens zur Quantenoptik. *PhyDid B—Didaktik der Physik—Beiträge zur DPG-Frühjahrstagung* **2020**, *Bonn*, 149–156. Available online: https://ojs.dpg-physik.de/index.php/phydid-b/article/view/1022 (accessed on 15 September 2022).
64. Scholz, R.; Wessnigk, S.; Weber, K.A. A classical to quantum transition via key experiments. *Eur. J. Phys.* **2020**, *41*, 055304. [CrossRef]
65. Kohnle, A.; Rizzoli, A. Interactive simulations for quantum key distribution. *Eur. J. Phys.* **2017**, *38*, 035403. [CrossRef]
66. Bloom, Y.; Fields, I.; Maslennikov, A.; Rozenman, G.G. Quantum cryptography—A simplified undergraduate experiment and simulation. *Physics* **2022**, *4*, 104–123. [CrossRef]
67. Grangier, P.; Roger, G.; Aspect, A. Experimental evidence for a photon anticorrelation effect on a beam splitter: A new light on single-photon interferences. *Europhys. Lett.* **1986**, *1*, 173–179. [CrossRef]
68. Silberhorn, C. Detecting quantum light. *Contemp. Phys.* **2007**, *48*, 143–156. [CrossRef]
69. Kline, T. *Psychological Testing: A Practical Approach to Design and Evaluation*; SAGE Publications, Inc.: Thousand Oaks, CA, USA, 2005.
70. Jorion, N.; Gane, B.D.; James, K.; Schroeder, L.; DiBello, L.V.; Pellegrino, J.W. An analytic framework for evaluating the validity of concept inventory claims. *J. Eng. Educ.* **2015**, *104*, 454–496. [CrossRef]
71. Taber, K.S. The Use of Cronbach's alpha when developing and reporting research instruments in science education. *Res. Sci. Educ.* **2018**, *48*, 1273–1296. [CrossRef]
72. Cantó-Cerdán, M.; Cacho-Martínez, P.; Lara-Lacárcel, F.; García-Muñoz, Á. Rasch analysis for development and reduction of Symptom Questionnaire for Visual Dysfunctions (SQVD). *Sci. Rep.* **2021**, *11*, 14855. [CrossRef] [PubMed]
73. Winter-Hölzl, A.; Wäschle, K.; Wittwer, J.; Watermann, R.; Nückles, M. Entwicklung und Validierung eines Tests zur Erfassung des Genrewissens Studierender und Promovierender der Bildungswissenschaften. *Zeit. Pädag.* **2015**, *61*, 185–202. [CrossRef]
74. Planinic, M.; Boone, W.J.; Susac, A.; Ivanjek, L. Rasch analysis in physics education research: Why measurement matters. *Phys. Rev. Phys. Educ. Res.* **2019**, *15*, 020111. [CrossRef]
75. Rizopoulos, D. Package 'ltm'. 18 February 2022. Available online: https://cran.r-project.org/web/packages/ltm/ltm.pdf (accessed on 15 September 2022).
76. Robitzsch, A.; Kiefer, A.; Wu, M. Package 'TAM'. 28 August 2022. Available online: https://cran.r-project.org/web/packages/TAM/TAM.pdf (accessed on 15 September 2022).
77. Mair, P.; Hatzinger, R.; Maier, M.J.; Rusch, T.; Debelak, R. Package 'eRm'. 15 February 2022. Available online: https://cran.r-project.org/web/packages/eRm/eRm.pdf (accessed on 15 September 2022).
78. Matejak Cvenic, K.; Planinic, M.; Susac, A.; Ivanjek, L.; Jelicic, K.; Hopf, M. Development and validation of the Conceptual Survey on Wave Optics. *Phys. Rev. Phys. Educ. Res.* **2022**, *18*, 010103. [CrossRef]
79. Chen, W.-H.; Thissen, D. Local Dependence indexes for item pairs using item response theory. *J. Educ. Behav. Stat.* **1997**, *22*, 265–289. [CrossRef]
80. Christensen, K.B.; Makransky, G.; Horton, M. Critical values for Yen's Q_3 : Identification of local dependence in the Rasch model using residual correlations. *App. Psych. Meas.* **2017**, *41*, 178–194. [CrossRef]
81. Jang, E.E.; Roussos, L. An Investigation into the dimensionality of TOEFL using conditional covariance-based nonparametric approach. *J. Educ. Meas.* **2007**, *44*, 1–21. [CrossRef]
82. Susac, A.; Planinic, M.; Klemencic, D.; Milin Sipus, Z. Using the Rasch model to analyze the test of understanding of vectors. *Phys. Rev. Phys. Educ. Res.* **2018**, *14*, 023101. [CrossRef]
83. Jones, D.G.C. Teaching modern physics-misconceptions of the photon that can damage understanding. *Phys. Educ.* **1991**, *26*, 93–98. [CrossRef]
84. Kidd, R.; Ardini, J.; Anton, A. Evolution of the modern photon. *Am. J. Phys.* **1989**, *57*, 27–35. [CrossRef]
85. Nousiainen, M.; Koponen, I.T. Pre-service teachers' declarative knowledge of wave-particle dualism of electrons and photons: Finding lexicons by using network analysis. *Educ. Sci.* **2020**, *10*, 76. [CrossRef]
86. Hoehn, J.R.; Gifford, J.D.; Finkelstein, N.D. Investigating the dynamics of ontological reasoning across contexts in quantum physics. *Phys. Rev. Phys. Educ. Res.* **2019**, *15*, 010124. [CrossRef]
87. Kalkanis, G.; Hadzidaki, P.; Stavrou, D. An instructional model for a radical conceptual change towards quantum mechanics concepts. *Sci. Educ.* **2003**, *87*, 257–280. [CrossRef]
88. Veith, J.M.; Girnat, B.; Bitzenbauer, P. The role of affective learner characteristics for learning about abstract algebra: A multiple linear regression analysis. *EURASIA J. Math. Sci. Technol. Educ.* **2022**, *18*, em2157. [CrossRef]
89. Rasch, G. On general laws and the meaning of measurement in psychology. In *Proceedings of the Fourth Berkeley Symposium on Mathematical Statistics and Probability. Volume 4: Contributions to Biology and Problems of Medicine;, Berkeley, CA, USA, 20 June–30 July 1960*; Neyman, J., Ed.; University of California Press: Berkeley/Los Angeles, CA, USA, 1961; pp. 321–333. Available online: https://projecteuclid.org/Proceedings/berkeley-symposium-on-mathematical-statistics-and-probability/proceedings-of-the-fourth-berkeley-symposium-on-mathematical-statistics-and/toc/bsmsp/1200512872 (accessed on 15 September 2022).
90. Pant, H.A.; Tiffin-Richards, S.P.; Köller, O. Standard-Setting für Kompetenztests im Large-Scale-Assessment. Projekt Standardsetting. *Zeit. Pädag.* **2010**, *56*, 175–188. [CrossRef]
91. Duncan, R.G.; Hmelo-Silver, C.E. Learning progressions: Aligning curriculum, instruction, and assessment. *J. Res. Sci. Teach.* **2009**, *46*, 606–609. [CrossRef]

92. Wright, B.D.; Masters, G.N. *Rating Scale Analysis*; MESA Press: Chicago, IL, USA, 1982.
93. Prasad, S.; Scully, M.O.; Martienssen, W. A quantum description of the beam splitter. *Opt. Comm.* **1987**, *62*, 139–145. [CrossRef]
94. Loudon, R. *The Quantum Theory of Light*; Oxford University Press: Oxford, UK, 2000.
95. Cova, S.; Longoni, A.; Andreoni, A. Towards picosecond resolution with single-photon avalanche diodes. *Rev. Sci. Instr.* **1981**, *52*, 408–412. [CrossRef]
96. Cova, S.; Ghioni, M.; Lacaita, A.; Samori, C.; Zappa, F. Avalanche photodiodes and quenching circuits for single-photon detection. *Appl. Opt.* **1996**, *35*, 1956–1976. [CrossRef] [PubMed]
97. Zambra, G.; Andreoni, A.; Bondani, M.; Gramegna, M.; Genovese, M.; Brida, G.; Rossi, A.; Paris, M.G.A. Experimental reconstruction of photon statistics without photon counting. *Phys. Rev. Lett.* **2005**, *95*, 063602. [CrossRef] [PubMed]
98. Hong, C.K.; Mandel, L. Theory of parametric frequency down conversion of light. *Phys. Rev. A* **1985**, *31*, 2409–2418. [CrossRef] [PubMed]
99. Christ, A.; Fedrizzi, A.; Hüel, H.; Jennewein, T.; Silberhorn, C. Parametric down-conversion. In *Single-Photon Generation and Detection. Physics and Applications*; Migdall, A., Polyakov, S.V., Fan, J., Bienfang, J.C., Eds.; Elsevier: Amsterdam, The Netherlands, 2013; pp. 351–410. [CrossRef]
100. Christ, A.; Eckstein, A.; Mosley, P.J.; Silberhorn, C. Pure single photon generation by type-I PDC with backward-wave amplification. In *Proceedings of the European Conference on Lasers and Electro-Optics and the European Quantum Electronics Conference: CLEO Europe/EQEC 2009 (Abstracts), Munich, Germany, 14–19 June 2009*; IEEE: Piscataway, NJ, USA, 2009; p. 917. [CrossRef]
101. Couteau, C. Spontaneous parametric down-conversion. *Contemp. Phys.* **2018**, *59*, 291–304. [CrossRef]
102. Brida, G.; Degiovanni, I.P.; Genovese, M.; Migdall, A.; Piacentini, F.; Polyakov, S.V.; Ruo Berchera, I. Experimental realization of a low-noise heralded single-photon source. *Opt. Express* **2011**, *19*, 1484–1492. [CrossRef] [PubMed]
103. Lounis, B.; Orrit, M. Single-photon sources. *Rep. Prog. Phys.* **2005**, *68*, 1129–1179. [CrossRef]
104. Krapick, S.; Herrmann, H.; Quiring, V.; Brecht, B.; Suche, H.; Silberhorn, C. An efficient integrated two-color source for heralded single photons. *New J. Phys.* **2013**, *15*, 033010. [CrossRef]
105. Genovese, M.; Gramegna, M.; Brida, G.; Bondani, M.; Zambra, G.; Andreoni, A.; Rossi, A.R.; Paris, M.G.A. Measuring the photon distribution with ON/OFF photodetectors. *Laser Phys.* **2006**, *16*, 385–392. [CrossRef]
106. Kimble, H.J.; Dagenais, M.; Mandel, L. Photon antibunching in resonance fluorescence. *Phys. Rev. Lett.* **1977**, *39*, 691–695. [CrossRef]
107. Hobson, A. There are no particles, there are only fields. *Am. J. Phys.* **2013**, *81*, 211–223. [CrossRef]
108. Strnad, J. Photons in introductory quantum physics. *Am. J. Phys.* **1986**, *54*, 650–652. [CrossRef]

physics

MDPI

Review

What Does the Curriculum Say? Review of the Particle Physics Content in 27 High-School Physics Curricula

Anja Kranjc Horvat [1,2,*], Jeff Wiener [1], Sascha Marc Schmeling [1] and Andreas Borowski [2]

1 European Organization for Nuclear Research (CERN), 1211 Geneva, Switzerland
2 Institute of Physics and Astronomy, University of Potsdam, 14482 Potsdam, Germany
* Correspondence: anja.horvat@cern.ch

Abstract: This international curricular review provides a structured overview of the particle physics content in 27 state, national, and international high-school physics curricula. The review was based on a coding manual that included 60 concepts that were identified as relevant for high-school particle physics education. Two types of curricula were reviewed, namely curricula with a dedicated particle physics chapter and curricula without a dedicated particle physics chapter. The results of the curricular review show that particle physics concepts are explicitly or implicitly present in all reviewed curricula. However, the number of particle physics concepts that are featured in a curriculum varies greatly across the reviewed curricula. We identified core particle physics concepts that can be found in most curricula. Here, elementary particles, fundamental interactions, and charges were identified as explicit particle physics concepts that are featured in more than half of the reviewed curricula either as content or context. Indeed, theoretical particle physics concepts are more prominent in high-school physics curricula than experimental particle physics concepts. Overall, this international curricular review provides the basis for future curricular development with respect to particle physics and suggests an increased inclusion of experimental particle physics concepts in high-school physics curricula.

Keywords: curricular review; particle physics; high-school education

1. Introduction

Spectacular phenomena and ground-breaking discoveries in particle physics often shake the media. One of the main keywords in particle physics research is CERN, the European Organization for Nuclear Research, as the largest particle physics research institution in the world. While CERN is mostly famous for its state-of-the-art particle physics research, one of its main missions is also education. Thus, CERN offers various educational programmes for research professionals, high-school students, and high-school teachers from around the world. However, offering programmes for an international selection of high-school students and teachers can be challenging, especially as the curricula of different countries can differ significantly. Therefore, a broader overview of the particle physics content in high-school curricula is needed to better support future educational programmes for high-school students and teachers.

Why should one include particle physics? First, phenomena in particle physics are spectacular. From high-energy particle collisions in detectors weighing several thousand tonnes, particle accelerators deep underground spanning across borders, to matter-antimatter annihilation, particle physics often attracts the attention of the media—and students. Indeed, students have been shown to be interested in spectacular phenomena of physics [1]. Second, particle physics is not limited to spectacular phenomena in big research laboratories. Various particle physics applications are present in everyday lives (e.g., medical imaging or radiation therapy). As such, particle physics shows a recent image of physics, which can help students to increase their knowledge of the nature of science made over the past few decades [2–7]. For example, one prominent call for including particle physics in more curricula comes directly from the latest European Strategy for Particle Physics Update, the cornerstone of the decision-making process in Europe for the long-term future of the field of particle physics, which states that: "The particle physics community should work with educators and relevant authorities to explore the adoption of basic knowledge of elementary particles and their interactions in the regular school curriculum." [8] (p. 13).

Since the first calls, several high-school physics curricula have already introduced particle physics. However, changes to curricula are slow [9,10] and most curricula still lag behind.

There are no research-based guidelines on how to introduce particle physics in high-school education. Therefore, an overview of practices in other educational systems can be beneficial. Indeed, curricular reviews are common practice in curriculum development (see, e.g., [11,12]). However, current reviews of particle physics content in high-school curricula are either limited to one educational system [6] or do not explicitly target particle physics [13]. Additionally, all efforts mostly focus on explicit particle physics in the curricula. Nevertheless, particle physics can be connected to other curricular topics as well. For example, when explaining the acceleration of particles in a particle accelerator, one can highlight aspects of mechanics, electromagnetism, quantum physics, and special relativity [14]. As such, neither a generalized overview of particle physics in high-school physics curricula nor high-school education in general can be inferred from the reviewed studies. Therefore, we designed a study to address this research gap by investigating the following questions:

1. Which concepts related to particle physics are the most common in national and international high-school physics curricula *with* a dedicated particle physics chapter?
2. Which concepts related to particle physics are the most common in national and international high-school physics curricula *without* a dedicated particle physics chapter?
3. What are the differences and similarities between particle physics contents in curricula with and without a dedicated particle physics chapter?

Overall, this study aims to give a clear overview of the particle physics content in various national and international curricula by taking stock of which particle physics concepts appear in the reviewed curricula.

2. State of the Research

This study reviewed particle physics content in high-school curricula. The following Sections focus on clarifying the necessary definitions and reviewing the current state of the research.

2.1. The Definition of a Curriculum

The term curriculum does not have a consensual definition in educational literature. The differences in the definition often stem from the various roles and characteristics that define curricula [15,16]. Generally, curriculum theorists distinguish three categories of curricula based on their role: formal, intended, and enacted curriculum [16–18]. This study focused on formal curricula. Formal curricula describe a set of guidelines for what should be taught in school and are designed and outlined by stakeholders of the respective educational system.

However, formal curricula can differ greatly depending on the educational system and the purpose of the curriculum in the respective system. Some curricula provide general guidelines with main key concepts (e.g., the Austrian curriculum), while others provide a detailed syllabus (e.g., the International Baccalaureate (IB) curriculum). Thus, curricula can be difficult to compare directly. An additional problem in curricular review is the language obstacle. Indeed, high-school curricula are typically written in the respective language of the instruction (e.g., the Slovenian curriculum is written in Slovenian). Furthermore, research suggests that curricula are subject to different interpretations [16].

Specifically, teachers interpret the formal curriculum within the scope of their teaching practices, which can lead to differences between the formal curriculum and the enacted curriculum. The possible differences in interpretation must be considered when designing a curriculum review.

2.2. Particle Physics in High-School Education

Particle physics is often perceived as a difficult topic in physics, associated with complex mathematics and difficult concepts. This perception can make the introduction of particle physics in high schools sound daunting and challenging. However, studies have shown that students can understand basic concepts of particle physics. For example, a study by Gourlay looked at high-school students' understanding of particle physics after they had explicit particle physics lessons [19]. The study showed that students could learn about quarks, leptons, particle systems, and annihilation. However, the students' knowledge was mostly declarative, which might reflect the curricular structure and content. For example, students knew that "electron is a type of lepton". Yet, they also mistakenly remembered matter as "made of particles and antiparticles" [19] (p. 6).

Similarly, Tuzón and Solbes reviewed the knowledge of Spanish high-school students that had no explicit lessons on particle physics before the study [5]. They had also shown some understanding of basic particle physics concepts. For example, half of the students knew that new particles can be observed from collisions. However, their overall particle physics knowledge was unstructured and incomplete. Indeed, without an explicit focus on particle physics, students only learn about particle physics implicitly and in fragments. Here, a structured introduction to particle physics can help students organize and expand their existing particle physics knowledge in a more meaningful way.

An introduction of particle physics as an explicit learning unit has not yet been empirically explored in the literature. Still, a few different approaches can be found. First, particle physics can be introduced through the perspective of particle interactions [20–22]. Here, particle interactions serve as a context for introducing the different charges, fundamental interactions, and elementary particles, which build the basis of the Standard Model of particle physics. Overall, this approach focuses on introducing mostly theoretical particle physics concepts.

Second, the challenge of teaching particle physics can be approached from the perspective of quantum mechanics [23,24]. This learning unit builds upon electromagnetism, quantum field theory, and quantum electrodynamics. However, students do not necessarily have experience with quantum field theory and quantum electrodynamics. While this approach does touch upon the reasons behind particles' existence, which is not the case in other approaches, its implementation might be too advanced and time-consuming to be included in high-school education. Furthermore, only theoretical particle physics concepts and some history of particle physics are included in this example.

Last, one suggestion features experimental particle physics concepts explicitly. Here, Polen proposes a broader chapter, including all concepts mentioned above with the addition of particle accelerators and detectors [25]. However, this suggestion is less detailed and provides less insight into the lesson structure. Thus, which concepts should be included and how they should connect within the lesson is unclear.

Likewise, suggestions for introducing individual particle physics concepts also focus more on theoretical particle physics than experimental particle physics. For example, several authors suggest using the Standard Model of particle physics as a possible window into the world of particle physics [22,26,27]. However, the Standard Model is a complex mathematical description of particle physics that goes beyond the level of high-school physics education. This level of complexity can lead to a naïve and reductionistic approach to introducing the Standard Model of particle physics [28,29]. Therefore, the complex mathematics of the Standard Model of particle physics is probably best avoided in the classroom. Here, for example, Lindenau and Kobel suggested a simplified introduction to the Standard Model of particle physics. Indeed, they focus on the three pillars of the

Standard Model of particle physics: elementary particles, fundamental interactions, and charges [22]. Furthermore, they connect these three concepts to other curricular topics to allow for better integration of particle physics.

Several other publications also explored the introduction of the three main concepts of the Standard Model of particle physics, i.e., elementary particles, fundamental interactions, and charges. Here, various innovative techniques have been suggested to enhance the introduction of elementary particles, for example, through hands-on activities and art [27,30–34]. For instance, McGinness et al. developed an activity using 3D-printed models of elementary particles, which students can use to learn about the different characteristics of elementary particles [31]. However, many of the published suggestions aim at introducing elementary particles and their properties in a declarative way, similar to the introduction of the periodic system of elements [20,22]. While students can memorize these concepts [19], it is questionable whether such rote learning contributes to the understanding of particle physics. One example that ties learning about elementary particles closer to understanding particle physics and how it is conducted are particle physics masterclasses. Here, students explore real data from particle colliders to identify which particles were created in a collision [35]. This approach is also one of the few that connect theoretical particle physics concepts, such as elementary particles and fundamental interactions, to experimental particle physics processes.

Several suggestions in the literature also explore the introduction of charges and fundamental interactions. Here, charges are typically mentioned as properties of elementary particles. In the example above, the 3D-printed elementary particle models by McGinness included information on particles' charges [31]. Likewise, the particle dance by Nikolopoulos and Pardalaki helped students embody elementary particles and particle interactions [27]. Moreover, the study by Wiener et al. that strongly focused on the representation of fundamental charges contextualized the concept of charges with elementary particles [36].

Overall, a big part of the reviewed suggestions on how to introduce particle physics in high-school education focused on theoretical particle physics concepts. Only a few suggestions for inclusions of experimental particle physics can also be found in the literature. Here, particle physics concepts are generally used as context for other curricular topics. For example, Barradas-Solas et al. suggested introducing particle detectors in the context of radiation [37]. Similarly, Cid-Vidal and Cid connected the acceleration of lead ions for particle collisions to electromagnetism [38]. Another extensive suggestion for contextualizing particle physics comes from Wiener et al. [14]. They suggested introducing particle physics concepts also in the contexts of mechanics, electromagnetism, thermodynamics, optics, radiation, quantum physics, and the theory of relativity. Indeed, experimental particle physics offers many links to other curricular topics. However, only a few mentioned publications on experimental particle physics explicitly included connections to theoretical particle physics concepts. For example, the above-mentioned suggestion by Wiener et al. connects charges and fundamental interactions to the Large Hadron Collider (LHC) at CERN, the largest particle accelerator in the world [14]. Additionally, Barradas-Solas [37] and Kvita et al. [39] use the context of particle detectors to introduce various elementary particles. Still, in most examples, experimental particle physics and theoretical particle physics concepts are presented as two separate entities with scarce connections.

Based on this literature review and the identified research gap, we conducted a curriculum review that is described in the following Sections.

3. Methods

This curriculum review consists of three steps. First, a coding manual was developed to guide the curriculum review. Second, the selected curricula were reviewed by experts. Last, the reviews were compared and analysed by the researchers. Each step is presented in more detail in the following Sections.

3.1. Coding Manual

The coding manual was developed following the dualistic method of deductive and inductive code creation [40,41]. First, the deductive part of the coding manual stemmed from previous research [11,42], including an expert concept mapping study with 13 experts in particle physics and physics education research [43]. Overall, the included concepts were not limited only to explicit particle physics, especially as curricula without a particle chapter were also included in the design. Indeed, concepts from other curricular topics were added to the review. These curricular topics were identified by the experts in the development phase of the coding manual as relevant for particle physics education, and they are also mentioned in various suggestions in the literature for particle physics introduction in high schools. For example, electromagnetism, special relativity, quantum physics, thermodynamics, radiation, and mechanics are some of the topics that were suggested by Wiener et al. as possible contexts for introducing particle physics concepts [14]. Second, the concepts that emerged from previous studies were used as the basis for an initial thematic analysis of a selected sample of curricula. Several additional concepts were added to the coding manual based on the patterns that emerged during this initial thematic analysis. For example, the initial analysis led to the addition of the concepts history of quantum physics and history of particle physics to the coding manual.

After the initial thematic analysis, the coding manual was re-tested on five curricula by at least two independent reviewers per curriculum. In addition to reviewing, the reviewers noted any ambiguities, problems, and suggestions they might encounter during their review. Next, the reviewers discussed their reviews with the first author. This discussion led to several modifications to the coding manual. For example, the concept electromagnetic waves was added to the coding manual due to its presence in several curricula and strong connection to particle accelerators. Additionally, the coding agreement increased from 60% to 100% after the review discussion.

Last, two physics education researchers and a particle physicist verified the updated coding manual. In the verification phase, some codes were combined based on their similarities. Indeed, various types of real-life applications (medical, technical) were merged into a bigger concept of real-life applications. Likewise, the concepts of Brout-Englert-Higgs field and Higgs boson were joined under the concept of Brout-Englert-Higgs mechanism. Additionally, the concept of Newtonian gravity was added to the category of other curricular topics. This addition ensured the separation between the concepts of Newtonian gravity and gravitational interaction. Furthermore, the concept of Einsteinian gravity was added to ensure that all mentions of gravity in the context of general relativity theory were coded separately from the concept of gravitational interaction in the context of particle physics. Likewise, the topic of particle transformations was split into three separate concepts: particle transformations, alpha radiation, and beta radiation. Here, the concepts of alpha radiation and beta radiation only included codes that appeared in the context of the topic of radiation. The concept of particle transformation only denoted processes of particle transformation in the explicit context of particle physics. Similarly, the concept of Feynman diagrams only included the explicit mention of Feynman diagrams as tools in particle physics. Furthermore, any indirectly hinted concepts (e.g., the concept of particle transformations is closely related to Feynman diagrams) were not to be coded unless explicitly mentioned elsewhere in the curriculum.

The final coding manual includes 60 codes that are either directly or indirectly related to particle physics. The codes are hierarchically organized into three levels. First, each code describes a specific concept, e.g., the concept of quarks. Second, the concepts are combined into topics, e.g., the concepts of quarks and leptons are in the topic of elementary particles. Overall, 25 topics were defined. Last, the topics are joined into three categories: explicit particle physics, other curricular topics, and history and nature of science. Within the category of explicit particle physics, some topics were additionally identified as theoretical or experimental particle physics topics based on their inherent nature. For example, the topic of elementary particles includes inherently theoretical particle physics concepts, while the topic of particle accelerators covers inherently experimental particle physics concepts. This distinction does not reflect the actual context in which the concepts appear in the curricula (e.g., the concept of leptons in an experimental context is still a theoretical particle physics concept).

Within the coding manual, each code is described by a short definition and a possible example from a curriculum, as shown in Table 1. Furthermore, exclusion criteria were added to codes that could be understood differently from what was intended (see the example of the concept of interaction particles in Table 1).

Table 1. Two examples of codes in the coding manual. Both examples fall into the category of explicit particle physics. The complete coding manual with all 60 codes can be found in the Supplementary Materials.

Topic	Code	Description	Exclusion	Example
Interaction particles	Interaction particles	The curriculum mentions bosons or at least one of the following: photons (as interaction particles), W bosons, Z bosons, gluons, gauge bosons.	To code the Higgs boson, please use the code "Brout-Englert-Higgs mechanism".	"The Standard Model explains three of the four (strong, weak and electromagnetic forces) in terms of an exchange of force-carrying particles called gauge bosons."
Elementary particles	Quarks	The curriculum mentions quarks or at least one of the following: up quark, down quark, strange quark, charm quark, top quark, bottom quark, anti-(up quark, down quark, etc.).		"Compare and contrast the up quark, the down quark, the electron and the electron neutrino, and their antiparticles, in terms of charge and energy (mass-energy)."

3.2. *Reviewing the Curricula*

The selection of the curricula was initially influenced by the fact that the study was conducted at CERN. Therefore, the initial selection of curricula was made by focusing on several CERN's member and associate member states. The list of all CERN's member and associate member states can be found on CERN's website [44]. The selection of countries was later expanded to include a more representative sample. Specifically, curricula from various other countries, such as Australia, Brazil, Russia, South Africa, and the United States of America, were added to the selection. Furthermore, the IB curriculum was included in the review due to its international character and relevance.

Several of the selected countries have more than one high-school physics curriculum. In that case, a curriculum with the most advanced physics curriculum was chosen for the review. This selection was made as particle physics is more likely to be included in more advanced curricula. Additionally, the curricula were chosen in a way that would reflect high-school students' expected knowledge at the end of their high-school education (i.e., age 18–19). With this selection, the study aimed to cover all relevant years of high-school education in a respective educational system. Overall, 27 curricula were selected, including 6 state curricula, 20 national curricula, and 1 international curriculum.

As mentioned above, the review and the comparison of different national and international curricula presented several obstacles. Thus, special attention was given to the selection of suitable reviewers. The curricula were reviewed by experts, namely teachers with experience in the relevant curriculum, sufficient particle physics knowledge and knowledge of the respective language. Two independent experts reviewed each curriculum following the coding manual to increase the reliability of the coding. In addition, each curriculum was examined by the first author, some with the help of online translators. Additional peer validation was conducted by the second author.

All discrepancies between the reviews of the experts and the researcher were addressed either by the first author directly or through discussion with the respective experts. For example, if one expert missed coding "the beginning and evolution of the Universe" as the concept of Big Bang, the first author would assign the respective code. Another example would be that "describe elementary particles" was coded as the concept of Standard Model of particle physics by one expert and not the other. Here, the first author discussed the discrepancy with the experts to find a consensus. In this example, the concept of Standard Model of particle physics was agreed not to be coded for that curriculum.

3.3. Analysis

The curricula were split into two types for analysis: curricula with a dedicated particle physics chapter and curricula without a dedicated particle physics chapter. Here, a chapter is defined as a section in the curriculum with an explicit focus on particle physics and a telling header, e.g., "Particle physics" or "The Standard Model of particle physics". Several curricula included the particle physics focus in a broader "Modern physics" chapter. These curricula were also counted as curricula with a dedicated particle physics chapter. Curricula without an explicit focus on particle physics or with no chapter were counted as curricula without a dedicated particle physics chapter.

The next step in the analysis was determining which concepts and topics are the most common in each type of curricula. As mentioned above, the review included 60 concepts that were grouped into 25 topics. Furthermore, the topics were grouped into three categories, namely the categories of explicit particle physics, other curricular topics, and history and nature of science. The number of curricula in which a specific concept would be featured was determined as follows: for each curriculum in which a concept was identified one or more times, the count increased by one. For example, if the concept of magnetic field was identified twice in the Slovenian curriculum, the count only increased by one. The topics were counted slightly differently. Here, a topic is counted as having appeared in a curriculum if at least one concept from the topic was identified in the curriculum. For example, the topic of charges contains concepts of electric charge, weak charge, and strong charge. Thus, the topic of charges was counted to appear in a curriculum if at least one of the three concepts was coded in the curriculum (e.g., the concept of electric charge). If a concept is featured multiple times in one curriculum or various concepts within the topic were identified in one curriculum, the count was still only increased by one.

Based on the count in the previous step, the concepts and topics that were found in more than half of the curricula were identified. Identifying the most featured concepts and topics allowed for a better comparison between the particle physics content in curricula with a dedicated particle physics chapter and curricula without a dedicated particle physics chapter. The results of this analysis are presented in Section 4.

4. Results

4.1. Curriculum Types

This curricular review reviewed 27 state, national and international high-school physics curricula. First, the structure of the curricula was inspected to define chapters and identify whether they include a dedicated particle physics chapter.

The 12 curricula with a dedicated particle physics chapter were the curricula of the IB [45], Austria [46], Australia—Queensland [47], Croatia [48], Germany—Brandenburg [49], Israel [50], Russia [51], Switzerland—Nidwalden [52], Serbia [53], South Africa [54], Spain [55], and the United Kingdom—A levels [56]. In most of these curricula, particle physics was included as a dedicated chapter towards the end of the curriculum. Thus, students would typically explicitly learn particle physics at the end of their high-school education, typically between ages 17 and 19.

The 15 curricula without a dedicated particle physics chapter were curricula of Brazil—São Paolo [57], Canada—Manitoba [58], Germany—Baden-Württemberg [59], Germany—Saxony [60], France [61], Ghana [62], Greece [63], Italy [64], Lebanon [65], Netherlands [66],

Poland [67], Slovakia [68], Slovenia [69], Sweden [70] and the United States of America—Next Generation Science Standards (NGSS) [71]. In both types of curricula, all chapters included in the curriculum were reviewed for explicit and implicit particle physics concepts.

4.2. Results of the Analysis

The curricula were reviewed based on the coding manual with 60 hierarchically organized concepts, as described above. The concepts were organized into 25 topics that were grouped into three categories, namely the categories of explicit particle physics, other curricular topics, and history and nature of science. After reviewers identified the presence of individual concepts, the number of times individual concepts appeared in different curricula was counted. Special focus was given to determining which concepts appear in more than half of the curricula.

The curricula were grouped and compared based on whether they contained a dedicated particle physics chapter. First, the curricula with a dedicated particle physics chapter were analysed. Here, 35 out of 60 concepts were found in more than half of the curricula with a dedicated particle physics chapter. Furthermore, 19 out of 25 topics were identified in more than half of said curricula. Second, the curricula without a dedicated particle physics chapter were analysed. Overall, 27 out of 60 concepts were identified in more than half of the curricula. Furthermore, 14 out of 25 topics appear in over half of the curricula without a dedicated particle physics chapter. The overall overlap between the curricula with and without a dedicated particle physics chapter is 80%. More details are presented in the paragraphs below.

4.2.1. Explicit Particle Physics

The results for the category of explicit particle physics are shown in Table 2 and Figure 1. Additionally, these results are summarized in the following paragraphs.

The review of the curricula with a dedicated particle physics chapter showed that the following nine explicit particle physics topics were identified in more than half of said curricula: topics of cosmology, the Standard Model of particle physics, fundamental interactions, charges, elementary particles, interaction particles, antimatter research, open questions in particle physics, and particle accelerators. All but two of these topics are from theoretical particle physics. All topics from the category of explicit particle physics were identified in at least one curriculum. Likewise, ten concepts were found in more than half of the reviewed curricula with a dedicated particle physics chapter: the concepts of Standard Model of particle physics, electromagnetic interaction, strong interaction, weak interaction, quarks, leptons, interaction particles, antimatter research, general particle accelerators, and open questions in particle physics. The following three concepts were not found in any reviewed curricula: the concepts of weak charge, strong charge, and modern particle detectors.

The review of the curricula without a dedicated particle physics chapter showed that four topics in the category of explicit particle physics were identified in more than half of the curricula, namely the topics of cosmology, fundamental interactions, charges, and elementary particles. Likewise, four concepts were found in more than half of the curricula, namely the concepts of Big Bang, electromagnetic interaction, electric charge, and leptons. Here, the review showed that the topic of cosmology and the concept of Big Bang typically appeared in parts of curricula that focused on astronomy. On the other hand, concepts within the topics of fundamental interactions, charges, and elementary particles were mostly featured in the topic of electromagnetism. They were represented by the concepts of electromagnetic interaction, electric charge, and an electron as an example of the concept of leptons, respectively.

Overall, the overlap of concepts appearing in curricula with and without a dedicated particle physics chapter is 32%. Each type of curricula includes concepts that the other type does not. The explicit particle physics concepts that appear in both types of curricula are the concepts of electromagnetic interaction, electric charge, and leptons. More than half of the curricula with a dedicated particle physics chapter also feature the concepts of the Standard Model of particle physics, strong interaction, weak interaction, quarks, interaction particles, antimatter research, general particle accelerators, and open questions in particle physics. Additionally, the following topics were identified only in more than half of the curricula with a dedicated particle physics chapter: the topics of the Standard Model of particle physics, interaction particles, particle accelerators, antimatter research, and open questions in particle physics.

Table 2. Overview of the concepts and topics within the category of explicit particle physics and their count across the curricula with and without a dedicated particle physics chapter. Topics in the first column are marked with T as theoretical and E as experimental.

Topic	Concept	12 Curricula *with* a Dedicated Particle Physics Chapter		15 Curricula *without* a Dedicated Particle Physics Chapter	
		Count (Topic)	Count (Concept)	Count (Topic)	Count (Concept)
1 T Cosmology	Big Bang	7 *	6	9 *	9 *
	Inflation		5		4
	Expansion		5		5
2 T Standard Model	Standard Model	8 *	8 *	3	3
3 T Fundamental interactions	Electromagnetic interaction	9 *	8 *	9 *	9 *
	Strong interaction		9 *		3
	Weak interaction		8 *		2
	Gravitational interaction		5		0
4 T Charges	Electric charge	11 *	11 *	11 *	11 *
	Strong charge		0		0
	Weak charge		0		0
5 T Elementary particles	Quarks	10 *	9 *	11 *	4
	Leptons		10 *		10 *
6 T Interaction particles	Interaction particles	8	8 *	7	7
7 T Brout-Englert-Higgs mechanism	Brout-Englert-Higgs mechanism	3	3	0	0
8 T Particle transformations	Particle transformations	6	6	4	4
9 T Feynman diagrams	Feynman diagrams	3	3	0	0
10 T Antimatter research	Antimatter research	7 *	7 *	4	4
11 E Particle accelerators	Linear accelerators	8 *	1	4	3
	Circular accelerators		5		3
	General particle accelerators		7 *		2
12 E Particle detectors	Historical detectors	4	1	2	1
	Modern detectors		0		0
	General particle detectors		3		1
13 E Data storage and data analysis	Data storage and data analysis	2	2	3	3
14 Advances in particle physics	Experimental results	7 *	3	1	1
	Open questions		7 *		0
15 Real-life applications of particle physics	Real-life applications of particle physics	6	6	5	5

* Concepts and topics that were identified in more than six curricula with a dedicated particle physics chapter or in more than eight curricula without a dedicated particle physics chapter. T Topics that were categorized as theoretical particle physics topics. E Topics that were categorized as experimental particle physics topics.

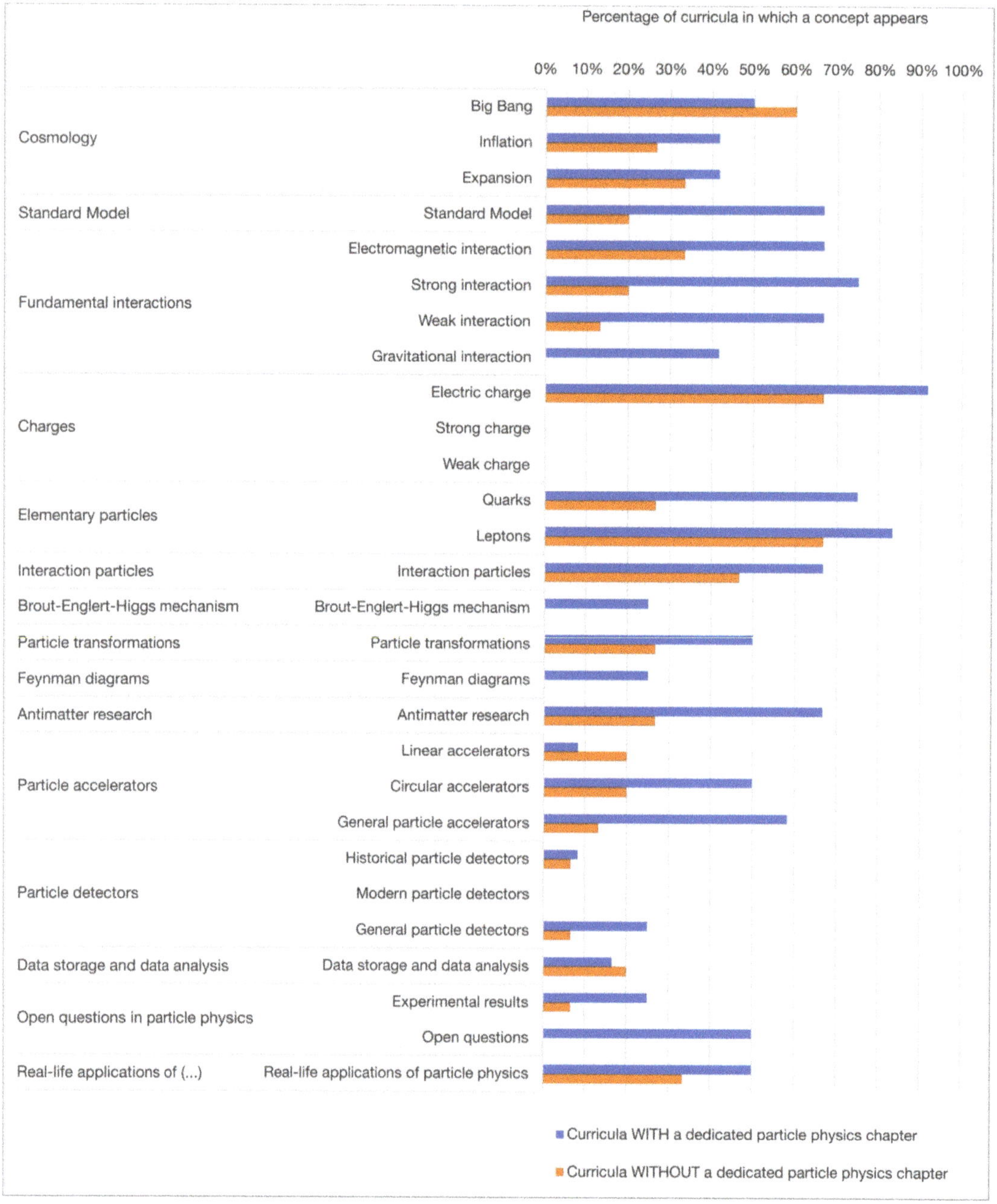

Figure 1. Graphic overview of the concepts within the category of explicit particle physics and the percentage of curricula in which they are featured. The bars in blue represent curricula with a dedicated particle physics chapter. The bars in orange represent curricula without a dedicated particle physics chapter. The figure corresponds to Table 2.

4.2.2. Other Curricular Topics

The results for the category of other curricular topics are shown in Table 3 and Figure 2. Additionally, these results are summarized in this Section.

The review of the curricula with a dedicated particle physics chapter showed that all topics from the category of other curricular topics are featured in all reviewed curricula. Similarly, 22 out of 29 concepts from the category of other curricular topics were identified in more than half of the curricula with a dedicated particle physics chapter. Only a few concepts were not found in more than half of the curricula, namely the concepts of Einsteinian gravity, conservation of angular momentum, conservation of charges, ionization, superconductivity, cosmic radiation, and quantum effects.

Table 3. Overview of concepts and topics within the category of other curricular topics and their count across the curricula with and without a dedicated particle physics chapter.

		12 Curricula *with* a Dedicated Particle Physics Chapter		**15 Curricula *without* a Dedicated Particle Physics Chapter**	
Topic	**Concept**	**Count (Topic)**	**Count (Concept)**	**Count (Topic)**	**Count (Concept)**
Mechanics	Linear motion	12 *	11 *	15 *	15 *
	Circular motion		11 *		15 *
Gravity	Newtonian gravity	12 *	12*	15 *	15 *
	Einsteinian gravity		5		2
Conservation Laws (of)	Energy	12 *	12 *	15 *	15 *
	Linear momentum		11 *		13 *
	Angular momentum		4		7
	Charges		6		4
Thermodynamics	Particle model	12 *	11 *	14 *	14 *
	Phase transitions		7 *		*
	Vacuum		7 *		4
Electromagnetism	Electric fields	12 *	12 *	15 *	14 *
	Magnetic fields		12 *		1 *
	Magnetic force		11 *		12 *
	Ionisation		5		10 *
	Electromagnetic waves		12 *		13 *
	Superconductivity		5		2
Radiation	Cosmic radiation	12 *	4	14*	3
	Alpha radiation		10 *		12 *
	Beta radiation		10 *		12 *
	Gamma radiation		10 *		12 *
	Radiation (general)		12 *		14 *
Special relativity	Relativistic motion	12 *	11 *	11 *	7
	$E = mc^2$		11 *		11 *
Quantum physics	Quantum effects	12 *	2	15 *	4
	Probability in quantum physics		11 *		11 *
	Atomic models		12 *		14*
	Atomic energy levels		11 *		11 *
	Quantum mechanics		10 *		9 *

* Concepts and topics that were identified in more than six curricula with a dedicated particle physics chapter or in more than eight curricula without a dedicated particle physics chapter.

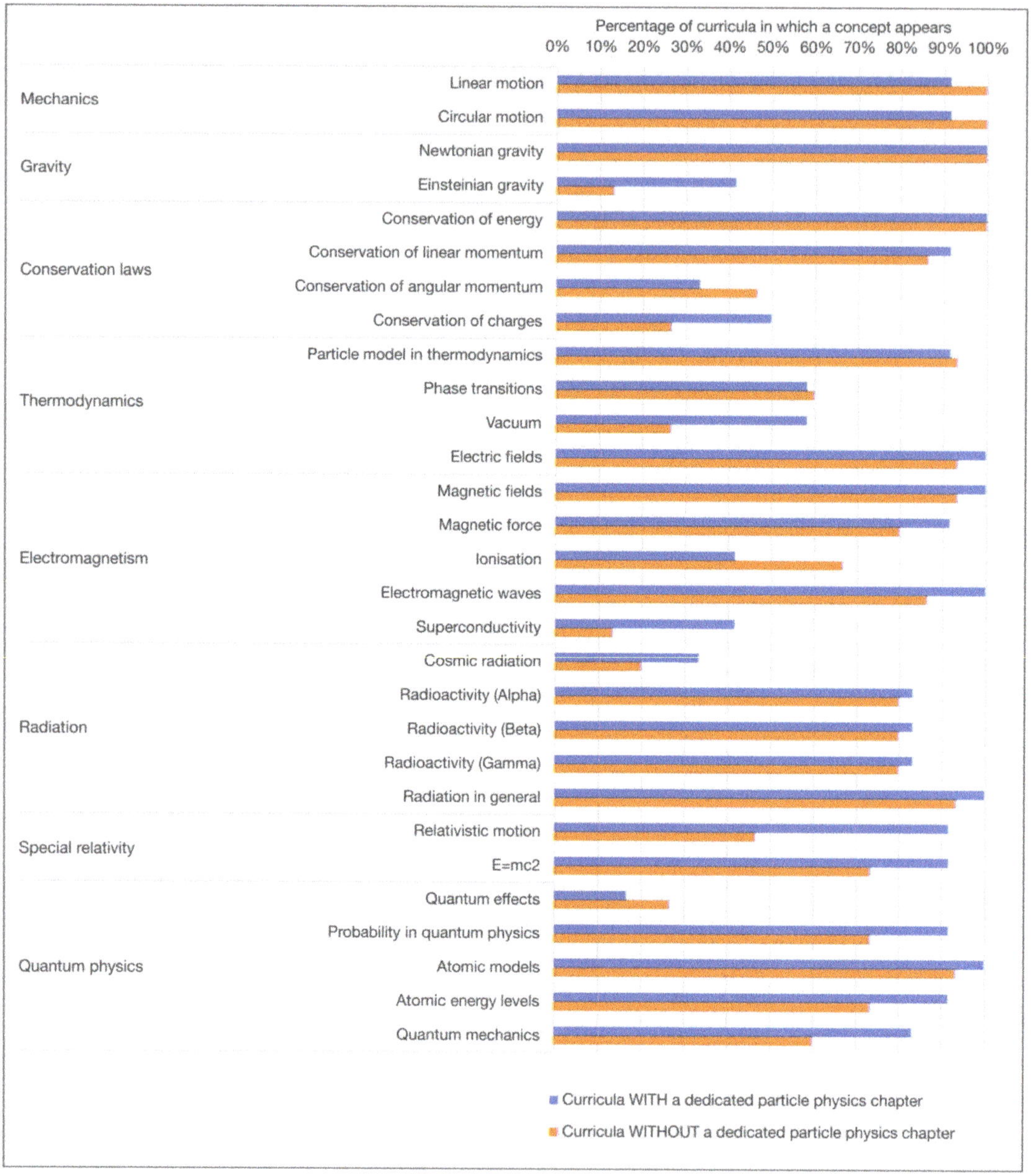

Figure 2. Graphic overview of the concepts within the category of other curricular topics and the percentage of curricula in which they are featured. The bars in blue represent curricula with a dedicated particle physics chapter. The bars in orange represent curricula without a dedicated particle physics chapter. The figure corresponds to Table 3.

Likewise, all topics from the category of other curricular topics are featured in more than half of the curricula without a dedicated particle physics chapter. Only the topics of thermodynamics, radiation, and special relativity were not identified in all curricula. Similarly, 21 out of 29 concepts from the category of other curricular topics were identified in more than half of the curricula without a dedicated particle physics chapter. Last, nine concepts were not found in more than half of the curricula, namely the concepts of Einsteinian gravity, conservation of angular momentum, conservation of charge, vacuum, superconductivity, cosmic radiation, relativistic motion, and quantum effects.

A 90% overlap was identified between the two types of curricula. Concepts that were identified in more than half of the curricula with a dedicated particle physics chapter and not in the other type are the concepts of vacuum and relativistic motion. Conversely, curricula without a dedicated particle physics chapter also feature the concept of ionization, which is typically introduced in the context of radiation.

4.2.3. History and Nature of Science

The results for the category of history and nature of science are shown in Table 4 and Figure 3. The category of history and nature of science has a complete overlap (100%) between the two types of curricula. Out of the three concepts in the category of history and nature of science, two were featured in more than half of the curricula: the concepts of history of quantum physics and the nature of science. Here, the concept of nature of science was typically present only implicitly. Most curricula listed aspects of the nature of science as general goals of physics as a subject. However, most curricula explicitly mentioned the topic of history of science within the content part of the curricula (either as content or context).

Table 4. Overview of concepts and topics within the category of history and nature of science and their count across the curricula with and without a dedicated particle physics chapter.

		12 Curricula *with* a Dedicated Particle Physics Chapter		**15 Curricula *without* a Dedicated Particle Physics Chapter**	
Topic	**Concept**	**Count (Topic)**	**Count (Concept)**	**Count (Topic)**	**Count (Concept)**
History of science	History of quantum physics	10 *	10 *	10 *	10 *
	History of particle physics		3		2
Nature of science	Nature of science	10 *	10 *	13 *	13 *

* Concepts and topics that were identified in more than six curricula with a dedicated particle physics chapter or in more than eight curricula without a dedicated particle physics chapter.

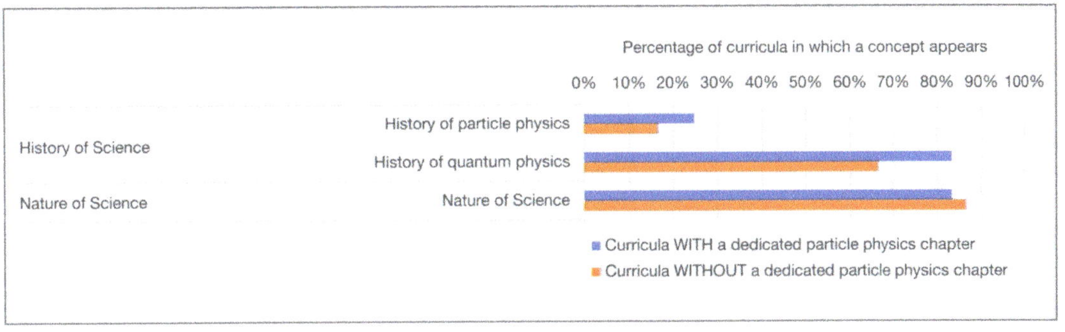

Figure 3. Graphic overview of the concepts within the category of history and nature of science and the percentage of curricula in which they are featured. The bars in blue represent curricula with a dedicated particle physics chapter. The bars in orange represent curricula without a dedicated particle physics chapter. The figure corresponds to Table 4.

5. Discussion

This curricular review investigated the particle physics content in 27 state, national, and international high-school physics curricula. Two types of curricula were reviewed: namely curricula with a dedicated particle physics chapter and curricula without a dedicated particle physics chapter. The analysis resulted in an overview of particle physics topics and concepts that are featured in more than half of the curricula with or without a dedicated particle physics chapter.

The reviewed curricula have many similarities. First, both types of curricula have a stronger focus on theoretical than experimental particle physics. Indeed, only curricula with a dedicated particle physics chapter feature experimental particle physics concepts. This result corresponds to what was found in the literature. Indeed, most suggestions for the introduction of particle physics focused more on theoretical than experimental particle physics concepts, e.g., [20–24]. Furthermore, the concepts that were most prominently mentioned in the literature are also present in more than half of the reviewed curricula, namely the topics of fundamental interactions, elementary particles, and charges. The results of this curricular review show that curricula with a dedicated particle physics chapter come somewhat close to Polen's suggested learning unit [25]. They suggested a learning unit that would include understanding particle accelerators and particle detectors as one of the goals. Indeed, most reviewed curricula also mention particle accelerators, at least as context for electromagnetism or circular motion. However, the topic of particle detectors is not present in a great majority of the curricula, regardless of how much the curriculum focuses on particle physics. Moreover, none of the curricula mentioned the concept of modern particle detectors. In the literature, particle detectors are generally introduced as a context for other curricular topics, for example, the topic of radiation—if at all [37]. Here, not including particle detectors appears to be a lost opportunity for more active learning and better representations. Indeed, introducing particle detectors in the classroom can help students better visualize an otherwise very abstract chapter [72].

The lack of including experimental aspects is not limited to particle physics. While experiments play a central role in physics education, their importance and value are often not introduced well enough in the physics education literature [73,74]. Specifically, Park et al. noted that physics is mostly presented in curricula and textbooks as "hardened facts with its making process erased" [75] (p. 1078). Thus, students often struggle to understand the role of experiments in science [75,76]. For example, most students think that confirming previously known results is the main goal of doing experiments [75]. Explicitly including aspects of modern experimental physics in curricula could help change this conception.

Second, three explicit particle physics concepts were not identified in any curriculum: the concepts of strong charge, weak charge, and modern particle detectors. Indeed, while electric charge has been identified as one of the most common concepts, the concepts of strong charge and weak charge are not mentioned in any curricula. However, the concept of electric charge was commonly mentioned in the context of electromagnetism, where it was typically referred to only as merely "charge". This nomenclature is not surprising, as no other charge appeared in the curriculum. Using "charge" to denote only electric charge also means that the concept of conservation of charge only constitutes the conservation of electric charge. However, the concepts of strong interaction and weak interaction are both mentioned in several curricula. Their description is generally limited to their role in nuclear reactions. Indeed, other studies have found that teachers often mention the concepts of strong interaction and weak interaction as decontextualized and in passing [5,77]. While the introduction of the strong interaction and weak interaction can include a short discussion about the nature of these fundamental interactions (e.g., range and relative strength), it typically avoids the associated charges. For example, the international baccalaureate (IB) curriculum, one of the most comprehensive curricula in this curricular review, prominently includes strong interaction and weak interaction [45]. However, while it describes the electromagnetic charge and the baryon numbers of elementary particles and interaction particles, the IB curriculum does not introduce strong charge or weak charge. Omitting the topic of charges from curricula means omitting the information about the role of charges in weak interaction and strong interaction. It can be argued that charges, especially strong charge and weak charge, can be a complex topic in particle physics. At higher levels of education, the topic of charges is often introduced through complex mathematics and difficult equations. However, as Wiener et al. [36] argued, the notion of charge can be introduced to learners as young as 12 years old without much complex mathematics. Furthermore, introducing at least the concepts of strong charge and weak charge can aid later studies by expanding the concept of charge to fields beyond electromagnetism.

Last, most curricula include the topic of history and nature of science, albeit most of them only implicitly. For example, the Austrian curriculum provides several aspects of nature of science at the beginning of the curriculum to highlight what teachers should convey throughout their teaching. However, the nature of science aspects are later not present in the content part of the Austrian curriculum. Indeed, the concept of nature of science is rarely found within the content part of curricula. Likewise, most textbooks only implicitly include the nature of science [78,79]. Yet, research shows that the nature of science is most effectively taught explicitly [80]. Therefore, finding appropriate context to support the teaching of the nature of science as content is crucial for teachers. Previous studies have suggested that contemporary topics can be used as the basis for teaching the nature of science [74,81]. Thus, as a contemporary field of physics, particle physics can help contextualize various aspects of the nature of science [82,83]. Here, the introduction of the concept of nature of science again calls for more experimental particle physics to be included in high-school education. Indeed, through learning about modern particle physics methods, students experience particle physics as "science-in-the-making". Through it, students can understand the principles of modern scientific explorations, from the importance of theoretical prediction to inference and social aspects of science. The role of theoretical predictions in designing future experiments and the tentativeness of science can further be highlighted through the discussion about open questions in particle physics. Indeed, open questions in particle physics can trigger conversations about science beyond what is typically perceived as facts. Likewise, by learning about real-life applications of particle physics, students can discover the value of modern science and its impact on life and society. As such, particle physics, similarly to other contemporary fields of science, presents itself as an excellent context for teaching the nature of science.

5.1. Strengths and Limitations

This international curricular review provides an overview of particle physics contents in 27 high-school state, national and international physics curricula. The curricula were screened using a coding manual that contained 60 concepts in particle physics and related topics. The coding manual was developed through an extensive literature review and the expert knowledge of multiple particle physicists and physics education researchers. Additionally, the manual was extensively pre-tested to ensure clarity. In its final form, the coding manual provides a clear and concise overview of particle physics concepts that experts perceive as the most relevant for high-school education. As such, the coding manual is a powerful tool both for evaluating particle physics content in curricula and possible further curricular development.

The curricular review was conducted by teachers with vast experience in the relevant curriculum, sufficient particle physics knowledge and knowledge of the respective language. These qualifications make them experts in their respective curricula, improving the quality of the review. Additionally, all curricula were reviewed by at least two experts in addition to being reviewed by the authors as well. The multitude of reviewers and the discussions to reach a consensus improved the overall reliability of the reviews.

However, the review of the curricula in this study was limited to science-oriented high-school physics curricula (when applicable) with no differentiation between elective and compulsory concepts. Such a selection of the curricula was purposeful to include the physics curricula with the maximum particle physics content in each country to determine the "ideal" scenario within physics curricula. However, this selection excludes curricula with less physics and curricula of other high-school subjects (e.g., astronomy and chemistry). This limitation can be addressed by broadening the scope of future studies to review curricula of other school types (e.g., technical high schools) and other school subjects.

Additionally, this curricular review focused on identifying which particle physics concepts are prominently featured in international curricula. As such, the review did not explore how particle physics concepts are introduced or how they are connected to other topics. Likewise, the levels of competence that students are expected to achieve in the context of particle physics were not analysed. The main reason for this decision is the stylistic differences between the curricula. Indeed, while some curricula are written as an extensive syllabus (e.g., the IB curriculum), others describe the required course content in open terms (e.g., the Austrian curriculum). However, the focus of this study was only to give an overview of which particle physics concepts are explicitly included in the curriculum. Therefore, the contextualization of particle physics and the level of competency students achieve on the topic should be a subject of a subsequent study.

Similarly, this curricular review focused on formal curricula provided by the governing bodies of the respective educational systems. Teachers, especially experienced teachers with deeper content knowledge in particle physics, can choose to include more concepts and more examples that are not mentioned in the formal curriculum. Likewise, how particle physics is introduced in textbooks can significantly change how particle physics is taught in classrooms. Indeed, textbooks are often used as a primary source of instruction [74]. As such, the results of this curricular review only show the baselines set by physics curricula. Further investigations into classroom practice and textbooks are needed to identify how teachers' instructional practices correspond to the official curriculum and how many elective topics are presented in the classroom. Furthermore, such investigations could provide a better overview of how particle physics context enters teachers' instructional practices.

5.2. Implications

The results of this first international curricular review of the particle physics content in high-school curricula show that particle physics is present in all high-school physics curricula, albeit not always explicitly. Regardless, particle physics is typically introduced with a strong focus on theoretical particle physics. Indeed, very little attention is given to experimental particle physics in high-school physics curricula. Based on these results, we can draw several implications for teaching, future curricular development, and future research, as presented below.

First, the coding manual developed for this curricular review is a powerful tool for identifying important concepts in particle physics education. The coding manual was based on the literature and further developed by experts in particle physics and physics education. Therefore, the manual itself can be used as a guide for teachers when implementing particle physics in their teaching practice.

Second, the curricular review uncovered many aspects for improving the curricula in general. The most striking is the lack of experimental particle physics concepts. Here, studies have shown that students often struggle with understanding the role of experiments in science [76,77]. Therefore, including more modern experimental physics in physics curricula can help students understand not only the importance but also the role of experiments in modern physics. Furthermore, several suggestions in the literature showed that experimental particle physics concepts could be a suitable context for other curricular topics, e.g., [14,37,38]. As such, connecting different parts of the curriculum to new concepts can help students to learn more meaningfully [84]. Additionally, giving a modern experimental example from an interesting topic in an otherwise classical topic can increase students' interest in other parts of physics as well. Indeed, especially real-life applications of experimental technologies in physics in fields such as medicine, food safety, and art (among others) can appeal to students on a personal level.

Likewise, particle physics can be a powerful context for learning about the nature of science. The results of this study show that the knowledge of the nature of science is often included as one of the overarching goals of physics education. Although the nature of science has been shown to be best taught explicitly [80], most curricula do not include it in the content part of the curricular document. Thus, a modern context such as particle physics can be used to introduce the nature of science more explicitly. However, further studies are needed to investigate how best to showcase the nature of science in the context of particle physics.

Last, the methodology in this curricular review relied greatly on expert practitioners, namely teachers with experience with their respective curricula. As such, the researchers were able to explore various curricula in different languages with little chance of mistranslation. Additionally, including several reviewers per curriculum reduced the possibility of misinterpretations of the curriculum and increased the reliability of the review. As such, the curriculum review was more streamlined, with less back-and-forth between the reviewers and the researchers. Therefore, this method is very suitable for future curriculum reviews.

To conclude, particle physics concepts are included in all curricula either explicitly or implicitly. Therefore, particle physics can provide students with an excellent example of science-in-action. However, particle physics is not the only modern physics topic that can fulfil that role. Indeed, quantum physics, nanotechnology, and soft condensed matter are only some examples of topics that can have a similar role in the curriculum. They are all examples of science-in-action, can be used as context for the explicit teaching of the Nature of Science, and are closely connected to various other parts of the curriculum. Furthermore, these topics might come even closer to students' everyday lives as they have or explicitly strive for more direct applications, e.g., nanotechnology for surface coatings or quantum computers. While there are several very prominent examples of particle physics applications in real-life situations (e.g., hadron therapy, positron emission tomography (PET), safety scanners), particle physics introduces a discussion on the value of purely basic science. Indeed, particle physics research is not explicitly aiming to result in a practical application, e.g., an equivalent of a quantum computer. The aim of particle physics remains to address the most fundamental questions of human nature: what are we made of, where do we come from, and where are we going. Therefore, the introduction of particle physics in high-school education provides a unique opportunity to discuss the importance of basic sciences in modern society.

Supplementary Materials: The following supporting information can be downloaded at: https://www.mdpi.com/article/10.3390/physics4040082/s1, Table S1: Coding Manual; Table S2: Overview of the particle physics topics and concepts in high-school curricula with and without a dedicated particle physics chapter.

Author Contributions: Conceptualization, A.K.H., J.W., S.M.S. and A.B.; methodology, A.K.H., J.W. and A.B.; validation, A.K.H. and J.W.; formal analysis, A.K.H.; investigation, A.K.H.; data curation, A.K.H.; writing—original draft preparation, A.K.H.; writing—review and editing, A.K.H., J.W. and A.B.; visualization, A.K.H.; supervision, J.W. and A.B.; funding acquisition, S.M.S. All authors have read and agreed to the published version of the manuscript.

Funding: This research received no external funding.

Data Availability Statement: The data presented in this study is presented in this article and its Supplementary Materials.

Acknowledgments: We thank all 64 teachers who helped with the reviewing of the curricula for providing their invaluable expertise during the review process.

Conflicts of Interest: The authors declare no conflict of interest.

References

1. Sjøberg, S.; Schreiner, C. Results and perspectives from the ROSE project: Attitudinal aspects of young people and science in a comparative perspective. In *Science Education Research and Practice in Europe*; Jorde, D., Dillon, J., Eds.; Brill: Leiden, The Netherlands, 2012; pp. 203–236. Available online: https://brill.com/view/book/edcoll/9789460919008/BP000010.xml (accessed on 27 September 2022).
2. Gil, D.; Solbes, J. The introduction of modern physics: Overcoming a deformed vision of science. *Int. J. Sci. Educ.* **1993**, *15*, 255–260. [CrossRef]
3. Chatterjee, L. Introducing particle physics through story. *Phys. Educ.* **2002**, *37*, 260–262. [CrossRef]
4. Kobel, M. High school students' exposure to modern particle physics. *Europhys. News* **2003**, *34*, 108–110. [CrossRef]
5. Tuzón, P.; Solbes, J. Particle physics in high school: A diagnose Study. *PLoS ONE* **2016**, *11*, e0156526. [CrossRef] [PubMed]
6. Hanley, P. Teaching particle physics. *Phys. Educ.* **2000**, *35*, 332–338. [CrossRef]
7. Aubrecht, G.J. Report on the conference on the teaching of modern physics. *Phys. Teach.* **1986**, *24*, 540–547. [CrossRef]
8. The European Strategy Group. *Deliberation Document on the 2020 Update of the European Strategy for Particle Physics*; Report CERN-ESU-014; CERN: Geneva, Switzerland, 2020. [CrossRef]
9. Young, M. What is a curriculum and whatcan it do? *Curric. J.* **2014**, *25*, 7–13. [CrossRef]
10. Fullan, M.G. Coordinating top-down and bottom-up strategies for educational reform. In *Systemic Reform. Perspectives on Personalizing Education*; Report No. OR 94-3308; Anson, R.J., Ed.; U.S. Department of Education, Office of Educational Research and Improvement: Washington, DC, USA, 1994; pp. 7–23. Available online: https://eric.ed.gov/?id=ED376557 (accessed on 27 September 2022).
11. Stadermann, H.K.E.; van den Berg, E.; Goedhart, M.J. Analysis of secondary school quantum physics curricula of 15 different countries: Different perspectives on a challenging topic. *Phys. Rev. Phys. Educ. Res.* **2019**, *15*, 010130. [CrossRef]
12. Lobato, T.; Greca, I.M. Análise da inserção de conteúdos de teoria quântica nos currículos de física do ensino médio. *Ciência Educ.* **2005**, *11*, 119–132. [CrossRef]
13. Mullis, I.V.S.; Martin, M.O.; Foy, P.; Hooper, M. *TIMSS Advanced 2015 International Results in Advanced Mathematics and Physics*; TIMSS & PIRLS International Study Center, Boston College: Chestnut Hill, MA, USA, 2016; Available online: http://timssandpirls.bc.edu/timss2015/international-results/advanced/ (accessed on 27 September 2022).
14. Wiener, G.J.; Woithe, J.; Brown, A.; Jende, K. Introducing the LHC in the classroom: An overview of education resources available. *Phys. Educ.* **2016**, *51*, 035001. [CrossRef]
15. Van den Akker, J. Curriculum perspectives: An introduction. In *Curriculum Landscapes and Trends*; Akker, J., Kuiper, W., Hameyer, U., Eds.; Springer Science+Business Meadia: Dordrecht, The Netherlands, 2003; pp. 1–10. [CrossRef]
16. Remillard, J.T. Examining key concepts in research on teachers' use of mathematics curricula. *Rev. Educ. Res.* **2005**, *75*, 211–246. [CrossRef]
17. Gehrke, N.J.; Knapp, M.S.; Sirotnik, K.A. Chapter 2: In search of the school curriculum. *Rev. Res. Educ.* **1992**, *18*, 51–110. [CrossRef]
18. Nawaz, H.; Akbar, R.A. Exploration of Gaps between intended and enacted physics curriculum: Teachers' professional development perspective. *Bull. Educ. Res.* **2019**, *41*, 1–10.
19. Gourlay, H. Learning about A level physics students' understandings of particle physics using concept mapping. *Phys. Educ.* **2017**, *52*, 014001. [CrossRef]
20. van den Berg, E.; Hoekzema, D. Teaching conservation laws, symmetries and elementary particles with fast feedback. *Phys. Educ.* **2006**, *41*, 47–56. [CrossRef]
21. Hoekzema, D.; Schooten, G.; van den Berg, E.; Lijnse, P. Conservation laws, symmetries, and elementary particles. *Phys. Teach.* **2005**, *43*, 266–271. [CrossRef]
22. Lindenau, P.; Kobel, M. Introducing an innovative approach of teaching the Standard Model of particle physics at high school. *J. Phys. Conf. Ser.* **2019**, *1287*, 012045. [CrossRef]
23. Hobson, A. Teaching elementary particle physics. Part I. *Phys. Teach.* **2011**, *49*, 12–15. [CrossRef]
24. Hobson, A. Teaching elementary particle physics. Part II. *Phys. Teach.* **2011**, *49*, 136–138. [CrossRef]
25. Polen, C. Particle physics: An essential and engaging part of the high school physics program. *Phys. Teach.* **2019**, *57*, 320–322. [CrossRef]
26. Woithe, J.; Wiener, G.J.; van der Veken, F.F. Let's have a coffee with the Standard Model of particle physics! *Phys. Educ.* **2017**, *52*, 034001. [CrossRef]
27. Nikolopoulos, K.; Pardalaki, M. Particle dance: Particle physics in the dance studio. *Phys. Educ.* **2020**, *55*, 025018. [CrossRef]
28. Michelini, M. Innovation of curriculum and frontiers of fundamental physics in secondary school: Research-based proposals. In *Fundamental Physics and Physics Education Research*; Sidharth, B.G., Murillo, J.C., Michelini, M., Perea, C., Eds.; Springer International Publishing: Cham, Switzerland, 2021; pp. 101–116. [CrossRef]
29. Passon, O.; Zügge, T.; Grebe-Ellis, J. Pitfalls in the teaching of elementary particle physics. *Phys. Educ.* **2019**, *54*, 015014. [CrossRef]
30. Alexopoulos, A.; Pavlidou, M.; Cherouvis, S. "Playing with Protons": A training course for primary school teachers at CERN. *Phys. Educ.* **2019**, *54*, 015013. [CrossRef]
31. McGinness, L.; Dührkoop, S.; Woithe, J.; Jansky, A. 3D printable quark puzzle: A model to build your own particle systems. *Phys. Teach.* **2019**, *57*, 526–528. [CrossRef]

32. MacDonald, T.; Bean, A. Quarked!—Adventures in particle physics education. *Phys. Teach.* **2009**, *47*, 38–41. [CrossRef]
33. Pavlidou, M.; Lazzeroni, C. Particle physics for primary schools—Enthusing future physicists. *Phys. Educ.* **2016**, *51*, 054003. [CrossRef]
34. Johansson, K.E.; Kobel, M.; Hillebrandt, D.; Engeln, K.; Euler, M. European particle physics masterclasses make students into scientists for a day. *Phys. Educ.* **2007**, *42*, 636. [CrossRef]
35. Cecire, K.; Bardeen, M.; McCauley, T. The CMS masterclass and particle physics outreach. *EPJ Web Conf.* **2014**, *71*, 00027. [CrossRef]
36. Wiener, G.J.; Schmeling, S.M.; Hopf, M. An alternative proposal for the graphical representation of anticolor charge. *Phys. Teach.* **2017**, *55*, 472–474. [CrossRef]
37. Barradas-Solas, F. Giving new life to old equipment. *Phys. Educ.* **2007**, *42*, 9–11. [CrossRef]
38. Cid-Vidal, X.; Cid, R. Lead ions and Coulomb's law at the LHC (CERN). *Phys. Educ.* **2018**, *53*, 024002. [CrossRef]
39. Kvita, J.; Čermáková, B.; Matulová, N.; Poštulka, J.; Staník, D. Particle camera MX-10 in physics education. *Phys. Educ.* **2019**, *54*, 025011. [CrossRef]
40. Fereday, J.; Muir-Cochrane, E. Demonstrating rigor using thematic analysis: A hybrid approach of inductive and deductive coding and theme development. *Int. J. Qualit. Meth.* **2006**, *5*, 80–92. [CrossRef]
41. Roberts, K.; Dowell, A.; Nie, J.-B. Attempting rigour and replicability in thematic analysis of qualitative research data; a case study of codebook development. *BMC Med. Res. Methodol.* **2019**, *19*, 66. [CrossRef] [PubMed]
42. Pavlin, J.; Stefanel, A.; Lindenau, P.; Kobel, M.; Kranjc Horvat, A.; Wiener, J.; Schmeling, S.; Borowski, A.; Sokołowska, D.; Čepič, M. Introduction of contemporary physics to pre-university education. In *Teaching-Learning Contemporary Physics. From Research to Practice*; Jarosievitz, B., Sükösd, C., Eds.; Springer International Publishing: Cham, Switzerland, 2021; pp. 71–90. [CrossRef]
43. CERN. Our Member States. Available online: https://home.cern/about/who-we-are/our-governance/member-states (accessed on 27 September 2022).
44. Kranjc Horvat, A. Particle Physics in High-School Education: What Should Students and Teachers Learn? Ph.D. Thesis, University of Potsdam, Potsdam, Germany, 2022.
45. International Baccalaureate Organization. Diploma Programme Physics Guide: First Assessment 2016. Available online: http://www.holyheart.ca/wp-content/uploads/2016/10/IB-Physics-Guide-2016.pdf (accessed on 27 September 2022).
46. Austrian Federal Ministry of Education, Science. Gesamte Rechtsvorschrift für Lehrpläne—Allgemeinbildende Höhere Schulen, Gymnasium Physik. 2018. Available online: https://www.ris.bka.gv.at/GeltendeFassung.wxe?Abfrage=Bundesnormen&Gesetzesnummer=10008568 (accessed on 27 September 2022). (In German)
47. Queensland Curriculum and Assessment Authority. Physics Geeneral Senior Syllabus 2019: Syllabus v1.3. Available online: https://www.qcaa.qld.edu.au/senior/senior-subjects/sciences/physics/syllabus (accessed on 27 September 2022).
48. Narodne Novine; Ministarstvo Znanosti i Obrazovanja. *Kurikulum Nastavnog Predmeta Fizika za Osnovne Škole i Gimnazije. Odluka, NN 10/2019-210.*. Available online: https://narodne-novine.nn.hr/clanci/sluzbeni/2019_01_10_210.html (accessed on 27 September 2022). (In Croatian)
49. *Rahmenlehrplan für den Unterricht in der Gymnasialen Oberstufe im Land Brandenburg. Physik*; Ministerium für Bildung, Jugend und Sport des Landes Brandenburg: Berlin–Brandenburg, Germany, 2018. Available online: https://bildungsserver.berlin-brandenburg.de/fileadmin/bbb/unterricht/rahmenlehrplaene/gymnasiale_oberstufe/curricula/2018/RLP_GOST_Physik_BB_2018.pdf (accessed on 27 September 2022). (In German)
50. Ministry of Education. Physics Curriculum for Upper Division. Available online: https://edu.gov.il/special/Curriculum/High-School/physics/Pages/Syllabus.aspx (accessed on 27 September 2022). (In Hebrew)
51. *Federal Component of the State Standard of General Education*; Ministry of Education of the Russian Federation: Moscow, Russia, 2016. Available online: https://mosmetod.ru/files/dokumenty/Primernaya-osnovnaya-obrazovatelnaya-programma-srednego-obshhego-obrazovaniya.pdf (accessed on 27 September 2022). (In Russian)
52. *Lehr-und Stoffplan*; Amt für Berufsbildung und Mittelschule, Kanton Nidwalden: Stans, Switzerland, 2020. Available online: https://www.kollegistans.ch/wp-content/uploads/LPSP-Kollegium-030522.pdf (accessed on 27 September 2022).
53. *Requirements for Curriculum and Program for Gift Students in Mathematics High School*; Ministry of Education, Science and Technological Development of Republic of Serbia: Belgrade, Serbia, 2016. Available online: https://www.mpn.gov.rs/wp-content/uploads/2018/02/%D0%9F%D1%80%D0%B0%D0%B2%D0%B8%D0%BB%D0%BD%D0%B8%D0%BA-%D0%BE-%D0%BD%D0%B0%D1%81%D1%82%D0%B0%D0%B2%D0%BD%D0%BE%D0%BC-%D0%BF%D0%BB%D0%B0%D0%BD%D1%83-%D0%B8-%D0%BF%D1%80%D0%BE%D0%B3%D1%80%D0%B0%D0%BC%D1%83-%D0%B7%D0%B0-%D0%BE%D0%B1%D0%B4%D0%B0%D1%80%D0%B5%D0%BD%D0%B5-%D1%83%D1%87%D0%B5%D0%BD%D0%B8%D0%BA%D0%B5-%D1%83-%D0%9C%D0%B0%D1%82%D0%B5%D0%BC%D0%B0%D1%82%D0%B8%D1%87%D0%BA%D0%BE%D1%98-%D0%B3%D0%B8%D0%BC%D0%BD%D0%B0%D0%B7%D0%B8%D1%98%D0%B8.pdf (accessed on 27 September 2022). (In Serbian)
54. Independent Examinations Board. *Advanced Programme Physics*; Independent Examinations Board: Gauteng, South Africa, 2019.
55. Ministerio de Educacion, Cultura y Deporte. I. Disposiciones Generales. *Bolet. Ofic. Estado* **2015**, *3/1*, 169–546. Available online: https://www.boe.es/boe/dias/2015/01/03/pdfs/BOE-A-2015-37.pdf (accessed on 27 September 2022). (In Spanish).
56. *AS and A-Level Physics*; AQA Education: Manchester, UK, 2017; Available online: https://filestore.aqa.org.uk/resources/physics/specifications/AQA-7407-7408-SP-2015.PDF (accessed on 27 September 2022).

57. *Currículo Paulista*; Governo do Estado de São Paolo, Secretaria da Educação: São Paolo, Brazil, 2021. Available online: https://efape.educacao.sp.gov.br/curriculopaulista/wpcontent/uploads/2021/01/TEMPLATE-Divis%C3%A3o-Habilidades-por-Ci%C3%AAncias-da-Natureza-e-suas-Tecnologias-1o-S%C3%A9rie-EM.pdf (accessed on 27 September 2022). (In Portuguese)
58. *Senior 3 Physics. A Foundation for Implementation*; Manitoba Departement of Education and Youth: Winnipeg, MB, Canada, 2003. Available online: https://www.edu.gov.mb.ca/k12/cur/science/found/physics30s/full_doc.pdf (accessed on 27 September 2022).
59. *Bildungsplan des Gymnasiums. Physik*; Ministerium für Kultus, Jugend und Sport Baden-Württemberg: Stuttgart, Germany, 2016. Available online: http://www.bildungsplaene-bw.de/site/bildungsplan/get/documents/lsbw/export-pdf/depot-pdf/ALLG/BP2016BW_ALLG_GYM_PH.pdf (accessed on 27 September 2022). (In German)
60. *Lehrplan Oberschule. Physik*; Sächsisches Staatsministeriums für Kultus: Dresden, Germany, 2019. Available online: http://lpdb.schule-sachsen.de/lpdb/web/downloads/54_lp_os_physik_2019.pdf?v2 (accessed on 27 September 2022). (In German)
61. Groupe Réussite. Programme de Physique-Chimie en Terminale. 2019. Available online: https://groupe-reussite.fr/ressources/programme-terminale-s-physique-chimie/ (accessed on 27 September 2022). (In French).
62. Ministry of Education. *Teaching Syllabus for Physics (Senior High School 1–3)*; Ministry of Education Science and Sports: Accra, Ghana, 2010. Available online: https://unesdoc.unesco.org/ark:/48223/pf0000221448.locale=en (accessed on 27 September 2022).
63. Ministry of Education and Religious Affairs; Institute of Educational Policy. Material and Instructions for the Teaching of Physical Sciences (Physics, Chemistry, Biology) in Grades A, B and C of the General High School. 2020. Available online: http://iep.edu.gr/el/graf-b-yliko/geniko-lykeio (accessed on 27 September 2022). (In Greek)
64. Ministero Dell'istruzione, Dell'università e della Ricerca. Indicazioni Nazionali Riguardanti Gli Obiettivi Specifici Di Apprendimento Concernenti Le Attività e Gli Insegnamenti Compresi Nei Piani Degli Studi Previsti per i Percorsi Liceali. 2010. Available online: https://www.gazzettaufficiale.it/gunewsletter/dettaglio.jsp?service=1&datagu=2010-12-14&task=dettaglio&numgu=291&redaz=010G0232&tmstp=1292405356450 (accessed on 27 September 2022). (In Italian)
65. Center of Educational Research and Development. *Curriculum Decree & General Target*; Center for Educational Research and Development: Beirut, Lebanon, 1997.
66. *Syllabus Centraal Examen 2021*; College voor Toetsen en Examens: Utrecht, The Netherlands, 2019; Available online: https://www.examenblad.nl/examenstof/syllabus-2021-natuurkunde-vwo/2021/f=/natuurkunde_3_versie_vwo_2021_av.pdf (accessed on 27 September 2022). (In Dutch)
67. Ministerstwo Edukacji Narodowej. Fizyka. Liceum/Technikum. 2018. Available online: https://podstawaprogramowa.pl/Liceum-technikum/Fizyka (accessed on 27 September 2022). (In Polish)
68. Štátny Inštitút Odborného Vzdelávania. Fyzika. 2019. Available online: https://siov.sk/wp-content/uploads/2019/06/B10_VO_ČaP_FY_CH_BI.pdf (accessed on 27 September 2022). (In Slovak)
69. *Curriculum: High School Physics*; Ministry of Science, Education and Sports of Republic of Slovenia: Ljubljana, Slovenia, 2015. Available online: http://eportal.mss.edus.si/msswww/programi2017/programi/media/pdf/un_gimnazija/2015/UN-FIZIKA-gimn-12.pdf (accessed on 27 September 2022). (In Slovenian)
70. Swedish National Agency for Education. Fysik. 2020. Available online: https://www.skolverket.se/undervisning/vuxenutbildningen/komvux-gymnasial/laroplan-for-vux-och-amnesplaner-for-komvux-gymnasial/amne?url=1530314731%2Fsyllabuscw%2Fjsp%2Fsubject.htm%3FsubjectCode%3DFYS%26tos%3Dvuxgy%26p%3Dp&sv.url=12.b173ee8160557dd0b8100d (accessed on 27 September 2022). (In Swedish)
71. National Academy of Sciences. Next Generation Science Standards. Available online: https://www.nextgenscience.org (accessed on 27 September 2022).
72. Keller, O.; Schmeling, S.; Müller, A.; Benoit, M. IPadPix—A novel educational tool to visualise radioactivity measured by a hybrid pixel detector. *J. Instrum.* **2016**, *11*, C11032. [CrossRef]
73. Koponen, I.T.; Mäntylä, T. Generative role of experiments in physics and in teaching physics: A suggestion for epistemological reconstruction. *Sci. Educ.* **2006**, *15*, 31–54. [CrossRef]
74. Park, W.; Yang, S.; Song, J. When modern physics meets nature of science: The representation of nature of science in general relativity in new Korean physics textbooks. *Sci. Educ.* **2019**, *28*, 1055–1083. [CrossRef]
75. Angell, C.; Henriksen, E.K.; Isnes, A. physics: Frightful, but fun. Pupils' and teachers' views of physics and physics teaching. *Sci. Educ.* **2004**, *88*, 683–706. [CrossRef]
76. Wilcox, B.R.; Lewandowski, H.J. Students' views about the nature of experimental physics. *Phys. Rev. Phys. Educ. Res.* **2017**, *13*, 020110. [CrossRef]
77. Sirola, C. May the forces be with you! *Phys. Teach.* **2018**, *56*, 118–119. [CrossRef]
78. Campanile, M.F.; Lederman, N.G.; Kampourakis, K. Mendelian genetics as a platform for teaching about nature of science and scientific inquiry: The value of textbooks. *Sci. Educ.* **2015**, *24*, 205–225. [CrossRef]
79. McDonald, C.V. Exploring representations of nature of science in Australian junior secondary school science textbooks: A case study of genetics. In *Representations of Nature of Science in School Science Textbooks*; McDonald, C.V., Abd-El-Khalick, F., Eds.; Routledge: New York, NY, USA, 2017; pp. 98–117.
80. Lederman, N.G. Nature of science: Past, present, and future. In *Handbook of Research on Science Education*; Abell, S.K., Lederman, N.G., Eds.; Routledge: New York, NY, USA, 2007; pp. 845–894.

81. Stadermann, H.K.E.; Goedhart, M.J. Why and how teachers use nature of science in teaching quantum physics: Research on the use of an ecological teaching intervention in upper secondary schools. *Phys. Rev. Phys. Educ. Res.* **2021**, *17*, 20132. [CrossRef]
82. Woithe, J.; Boselli, M.; Chatzidaki, P.; Dahlkemper, M.N.; Duggan, R.; Durey, G.; Herff, N.; Kranjc Horvat, A.; Molaro, D.; Scheerer, G.W.; et al. Higgs in a box: Investigating the nature of a scientific discovery. *arXiv* **2022**, arXiv:2206.06107. [CrossRef]
83. Kranjc Horvat, A.; Boselli, M.; Chatzidaki, P.; Dahlkemper, M.N.; Duggan, R.; Durey, G.; Herff, N.; Molaro, D.; Scheerer, G.W.; Schmeling, S.; et al. The Mystery Box Challenge: Exploring the Nature of Science. *Sci. Sch.* **2022**, *59*. Available online: https://www.scienceinschool.org/article/2022/mystery-box-challenge/ (accessed on 27 September 2022).
84. Ausubel, D.P. *The Acquisition and Retention of Knowledge: A Cognitive View*; Springer Media: Dordrecht, The Netherlands, 2000. [CrossRef]

physics

MDPI

Article

Opportunities and Challenges of Using Feynman Diagrams with Upper Secondary Students

Merten Nikolay Dahlkemper [1,2,*], Pascal Klein [2], Andreas Müller [3], Sascha Marc Schmeling [1] and Jeff Wiener [1]

1 European Organization for Nuclear Research (CERN), 1211 Geneva, Switzerland
2 Physics Education Research Section, Faculty of Physics, University of Göttingen, 37077 Göttingen, Germany
3 Faculty of Sciences, Section of Physics and University Institute of Teacher Education (IUFE), University of Geneva, 1211 Geneva, Switzerland
* Correspondence: merten.dahlkemper@cern.ch

Abstract: Particle physics is an exciting subject for high school students, and there have been various approaches on how to introduce the topic in the classroom. Feynman diagrams (FDs) are an often-used form of representation in particle physics and could play an important role in such an introduction. However, their potential educational value has not yet been investigated. To this end, we interviewed four experts in the field of particle physics education on the opportunities and challenges Feynman diagrams could pose for high school students. We analyzed their answers using a thematic analysis framework, categorizing them into five themes. The results of these interviews show that there are two challenges (FDs elicit and perpetuate inadequate conceptions about particle physics, and FDs can only be treated superficially in school) and three opportunities (FDs can link particle physics and other physics topics in high school education, FDs offer an opportunity for different particle physics topics to be taught, and FDs offer a connection to current research). The results of this expert interview study lead to several suggestions on how to design learning environments that incorporate Feynman diagrams.

Keywords: particle physics; Feynman diagrams; secondary education; expert interviews; design-based research

1. Introduction

Particle physics has been introduced in high school physics curricula in many countries. When teaching about particle physics, an often-used representation is that of so-called Feynman diagrams (FDs). However, while being of great use for particle physics, it is unclear whether using FDs in education benefits learners, since possible inadequate conceptions could be connected to them. The current study explores potential opportunities FDs can create for learning particle physics and the challenges that using these diagrams pose. To meet these aims, we interviewed experts in particle physics education.

1.1. Particle Physics in High School Education

The first calls to introduce particle physics in high school physics teaching date back to the 1980s [1], with one of the first conferences on the teaching of modern physics taking place at the European Organization for Nuclear Research (CERN) in 1984 [2]. Students might see particle physics as an exciting topic, which can serve as an example of the nature of physics as being tentative and constantly changing. Moreover, unknown phenomena and the newest scientific discoveries have been proven to be extremely interesting topics for adolescents [3]. Finding answers to the fundamental questions of nature can positively influence students' attitudes toward physics. The call for particle physics in high school education has since been replicated several times (e.g., [4–6]). Despite these calls, particle physics only rarely makes it into the curricula and subsequently into the teacher training courses at university. However, several suggestions have been made on how to incorporate particle physics into high school education (e.g., [7–10]). These suggestions vary greatly in the concepts they include and the approaches they take to introduce these concepts to high school students. They range from an overview of the "particle zoo" [10] to discussions of particle interactions, conservation laws, and symmetries [11,12] to coherent learning material for teaching the Standard Model of particle physics by connecting charges, particle interactions, and elementary particles [9].

Some approaches have been criticized as presenting the topic of particle physics in a superficial and oversimplified way [13,14]. Furthermore, presenting all the elementary particles in a merely enumerative manner might not be very motivating and might lead to rote learning of particle names instead of achieving a conceptual understanding of particle physics [15].

So far, few studies have been conducted to systematically investigate students' understanding of particle physics or the impact of specific interventions in the field [6,16,17].

1.2. Feynman Diagrams

A specific topic within particle physics is that of Feynman diagrams. The idea of FDs was first presented publicly by Richard Feynman at a conference in the spring of 1948 [18] and was subsequently published by Freeman Dyson and Feynman [19,20]. Since then, they have become a commonly used graphical tool in many areas of theoretical physics, but especially in quantum field theory and, thus, particle physics. In fact, they have been invented mainly as a tool for "bookkeeping" and visualization in the perturbation-theoretical calculation of processes in particle physics [18]. However, there is a debate about whether or not FDs are a form of representation of physical processes [21–23]. This debate dates to a disagreement between the inventor of the diagrams, Feynman, and the physicist who connected the diagram to a set of mathematical rules, Dyson. This disagreement is also known as the Feynman–Dyson split [18].

FDs are also part of many introductions to particle physics for high school students (e.g., [7–9]). However, Passon et al. [24] highlight various ideas associated with FDs that are present in educational material, which can be considered to be misleading. For example, any reading in which particles are assigned a trajectory in time and space is, according to the authors, physically nonsensical. Various physicists and physics educators have attempted to give educational introductions to FDs. For instance, Pascolini and Pietroni [25] introduced the diagrams as "accurate metaphors" and attempted to introduce the rules underlying Feynman diagrams to learners using a mechanical model. This approach was empirically investigated in a post-test one month after the intervention, where the learners performed satisfactorily. Hoekzema et al. [11,12] used a reduced form of FDs, which they call "reaction diagrams", to explain conservation laws and symmetries in particle physics. Their proposed teaching activity consists of two 50 min lessons embedded in a larger teaching project about nuclear reactions and elementary particles. The activity had been used for two years as part of a modern physics teaching project. In particular, the approach with the reduced form of the FDs was deemed more comprehensible than a previous text with conventional FDs.

Figure 1 shows a typical example of an FD. Explanations for this type of diagram differ in the extent to which it is explained 'literally'. "Literally" means, in this context, that the diagram is understood as a particle and an antiparticle traveling in straight lines toward each other, meeting and creating a photon, and, at some point, a new particle-antiparticle pair flying away from each other. While some parts of these explanations are compatible with the physical interpretation of the diagrams (for example, particles annihilate each other and new particles are created), others are rather inadequate conceptions (for example, particles do not have determined trajectories). In some texts, FDs are introduced as space–time diagrams [26], while in others, the individual fundamental vertices are explained and linked to mathematical terms in order to stress the inherent mathematical meaning of the diagrams [27]. Several authors try to avoid the problem of a too literal reading by emphasizing the fact that FDs contribute to a probability amplitude [7,28,29].

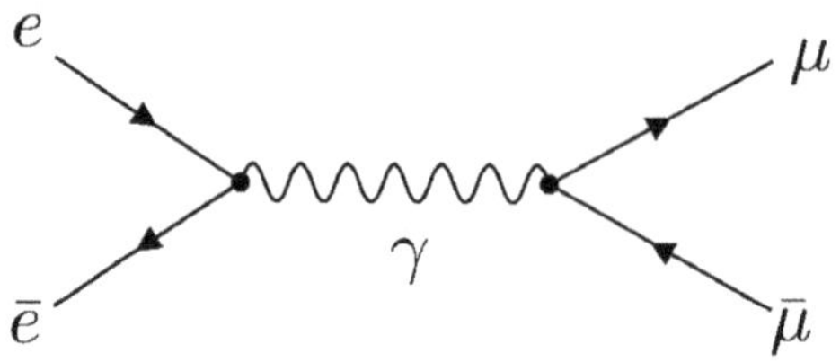

Figure 1. An example of a Feynman diagram of electron–positron ($e\bar{e}$) annihilation. The diagram describes the transformation of an electron–positron pair into a muon–antimuon ($\mu\bar{\mu}$) pair. The interaction is mediated by a photon (γ) which is the interaction particle of the electromagnetic interaction. The direction of arrows determines whether it is a particle or an antiparticle.

A defining property of every FD is that charges are conserved throughout the diagram, meaning that the total amount of any charge (electromagnetic, weak, and strong charge) is the same on the left and on the right side of the diagram. In Figure 1, this can be straightforwardly checked for the electric charge, since the electron and the positron on the left have a total electric charge of zero, which is also true for the muon and the antimuon on the right side.

Different tools have been developed to teach the underlying principles of FDs, most notably the conservation of charges. For example, *FeynGame* is a software that can be used to draw FDs and check them for correctness [30]. Another example is the educational card game *Feynman-Rhombino*, which allows players to create FDs by applying the rules of charge conservation [31].

1.3. Context of the Present Study: Design-Based Research

As illustrated above, several proposals suggest how FDs can be used for educational purposes. However, to our knowledge, no empirical study has investigated the suitability and feasibility of FDs for use in the physics classroom. Therefore, the overarching aim of this research project is to iteratively develop a learning environment that introduces FDs to upper secondary students (16–18-year-olds) and to test the educational use of Feynman diagrams. To do so, we use the framework of design-based research (DBR) [32].

DBR has been used to create learning environments in various fields of modern physics, for example, for the topics of chaotic systems [33], general relativity [34], special relativity [35], and, in particular, particle physics [17].

The DBR framework, also known as design research, design experiments, or educational research design, has been widely used and has resulted in learning interventions that create improved outcomes or student attitudes among the students [36,37]. There are many different approaches to DBR, but a central element is the iterative design and testing of the material. DBR produces both new theories and teaching practices to impact learning [38]. In this way, DBR is a method designed to bridge the gap between research and practice.

A particularly interesting link can be created between DBR and the model of educational reconstruction [39]. In this model, the educational design is informed both by the content structure of the subject matter and by the perspectives of the learners. Analysis of the content structure leads to the formulation of key messages, whereas that of the learners' perspective identifies relevant prior knowledge and potential learning difficulties of the respective subject matter. These two perspectives influence each other in an iterative process. Therefore, the reconstruction of the subject matter does not necessarily follow the content structure of the subject matter. Both the analysis of the content structure and the learners' perspective can be informed by different sources such as experts, literature, and the students themselves.

1.4. Scope of the Study

Building on the existing work about how (not) to incorporate FDs in educational material about particle physics, the current study presents a systematic analysis of the challenges and learning opportunities that are connected to the teaching of FDs. This analysis is the basis for an educational reconstruction of FDs, which is currently undergoing iterative testing with high school students. The research question that guided this work was: What challenges and opportunities do FDs pose for high school students?

2. Materials and Methods

To answer the research question, explorative interviews were conducted with experts in the field of particle physics education [40]. The aim was to determine and summarize the experts' opinions on what purpose FDs could serve in upper secondary school education.

2.1. Selection of Experts

The experts were selected based on their expertise in conveying particle physics to high school students.

We selected three researchers from two German universities who are involved in education programs in particle physics aimed at high school students. Two of them are education researchers, and one is a particle physicist. All three experts have previously published on the educational use of FDs.

After conducting interviews with these three experts and discussing the results among the research team, a discussion point regarding the epistemological role of FDs was raised. Therefore, a science philosopher was interviewed as fourth expert. This expert had expertise in the philosophy of science and conceptual analysis of contemporary physics.

The present study is exploratory in that, for the first time, it gathers data for the challenges and opportunities of integrating important content of contemporary physics into high school physics teaching. While the sample size is small, the results provide evidence for considerable and informative variance on the one hand and increasing overlap between the answer sets on the other hand. This is well in line with the main purpose of the present study, namely an in-depth insight about potential challenges and opportunities to guide further development, and not an exhaustive treatment of the topic (see [41,42]).

2.2. Conducting the Interviews

The first author conducted the interviews in a semi-structured way, that is, the interviewer had an interview guideline with the core questions, but he could diverge from it depending on the answers the interviewees gave.

The interview guideline was constructed based on two guiding questions, which resulted from the research question stated above. These were:

- What challenges are connected to teaching FDs to high school students?
- What opportunities for physics education at the high school level are provided by FDs?

The interview guideline, which is published in the Supplementary Material, contained 12 questions, which were divided into three parts: In the first part, the interviewees described what FDs are. This part served as an entry point to the interview and was not connected to one of the guiding questions. The second part was concerned with possible learning obstacles the experts could imagine or had encountered. This included inadequate conceptions students might already have, or what prior knowledge is needed to read and draw FDs. The third part was connected to the second guiding question and focused on potential learning opportunities offered by FDs. In this part, it was asked, for example, how an appropriate instruction would look like and how students could benefit from learning about Feynman diagrams. In the end, the experts were asked how they would describe a particular Feynman diagram to a high school student.

Special emphasis was put on appropriateness for the target group. For example, concerning the question of what a Feynman diagram is, the experts were specifically asked to describe it for a 17-year-old high school student. Similarly, when asked about potential learning obstacles, the experts were asked about learning obstacles specific to high school students.

The guideline for the interview with the fourth expert was adapted to his expertise, which differed from the expertise of the other interviewees. Specifically, the original guideline was shortened, and the following three introductory segments were added to the interview guideline to better align the results of the fourth interview with the first three interviews: The first segment was intended to clarify the definition of terms such as "model" and "theory". The second segment touched upon the connection between models and reality, since one of the open discussion points concerned the epistemological meaning of FDs. In this part, the expert was asked, for example, to which extent models play a role in science and what role mathematical descriptions play for models. The third segment focused on the role of modern physics in schools, since the second open discussion point after the first three interviews led to the question of whether modern physics—or particle physics in particular—should be taught in high school. In this part, the expert was asked, for example, which aspects of the nature of science he sees as vital for a meaningful discussion of science in a high school classroom.

The interviews were conducted in German via a videoconferencing software and were subsequently transcribed verbatim by the interviewer. The interviews were scheduled to take one hour and lasted for 63, 81, 37, and 81 min. The third expert mentioned during the interview that he did not feel completely well, which led to shorter answers and a faster interview pace, resulting in a shorter interview length.

2.3. Coding Scheme

The interview transcripts were categorized according to the thematic analysis framework [43]. The coding scheme contained two dimensions based on the guiding questions, that is, opportunities and challenges of FDs. These dimensions subsequently served as the highest level of the category system.

The transcripts were scanned for segments referring to these dimensions. Only those segments that answered one of the two guiding questions were categorized. Segments were of different lengths, depending on how much context was given within one segment. The shortest segments consisted of one sentence and the longest of about ten sentences. The first author inductively categorized these segments with preliminary category descriptions. The segments were categorized at the least general level possible, based on their thematic similarities. After categorizing approximately 30% of the transcripts, the categorization was discussed with the co-authors and revised, that is, categories were added, renamed, or merged. Based on the revised categorization, the transcripts were categorized in full. The categories were then subsumed into common themes and subthemes. Some subthemes also had sub-subthemes. Since, depending on the number of segments in a theme, the division into themes and (sub-)subthemes is different, the lowest level is generically called categories in the further description. Further on, categories were dissolved, and the segments were categorized into other existing categories. Through this process, the number of categories was reduced. This categorization was subsequently reviewed while the categories were given instructive descriptions to be incorporated into a coding manual. This coding manual was then discussed with two German-speaking doctoral students in the field of physics education who were not previously involved in the research process. Based on this discussion, the manual was again adapted by redefining category descriptions and merging categories to further reduce their number.

2.4. Validating the Coding of the Data

The revised coding manual was eventually given to the two doctoral students who used this coding manual to code approximately 10% of the coded text segments. This coding manual had 37 categories categorized into five themes. Interrating yielded a Krippendorff's alpha [44] of 0.49. This interrating was subsequently discussed between the first author and the two interraters. Interrater reliability after the discussion was increased to an acceptable level of 0.87. During this interrater discussion, several remarks about the coding manual were made, which led to another revision of both the coding manual and the coding of categories. Disagreements in the coding were caused by giving too little context within segments, which is why some segments were extended to give more context. The final coding manual is shown in the Supplementary Material. It is divided into two dimensions (challenges and opportunities), of which the first dimension contains 12 categories divided into two themes and four subthemes. The second dimension contains 12 categories divided into three themes. A second round of interrating was conducted with a new set of segments and a new interrater who had not been involved in the process before. This interrating showed an initial Krippendorff's alpha of 0.37, and after another interrater discussion and a revised coding, the interrater reliability showed 100% agreement.

3. Results

In total, five themes emerged from the analysis of all four expert interviews, which in turn were assigned to the two dimensions "challenges" and "opportunities". Two themes concerned challenges and three themes concerned opportunities. Of the challenges, the most prominent theme was that FDs elicit and perpetuate inadequate conceptions, which is in line with prior work (see Section 1.2). The second theme was that FDs can only be discussed superficially when discussing them with high school students. In contrast, among the opportunities, the following three themes became apparent: (a) FDs offer the opportunity to discuss perspectives on science, (b) FDs offer a link between particle physics and other high school physics topics, and (c) FDs can be a hook for different particle physics topics, such as the conservation of charges in particle transformation processes.

These five themes are explained further in the following text using subthemes and example segments from the transcripts. The segments presented here are translated from the original. The experts are referred here to as E1, E2, E3, and E4 in the following. E1, E2, and E3 are the first three experts, while E4 is the science philosopher who was later interviewed.

3.1. Challenges

Table 1 shows the two themes and corresponding subthemes present within the dimension of challenges. The subthemes are also further divided into sub-subthemes. Moreover, not all the themes were mentioned by all the experts. Table 1, therefore, shows which of the four experts mentioned which sub-subtheme.

Table 1. Themes and subthemes of the dimension challenges. The order of the (sub-)subthemes within a (sub)theme is according to the frequency of mentions.

Theme	Subtheme	Sub-Subtheme	Mentioned by Experts (E1, E2, etc.)
Feynman diagrams (FDs) elicit and perpetuate inadequate conceptions	Types of inadequate conceptions connected to FDs	Particle processes are embedded in spacetime	E1, E2, E3, E4
		FDs show observable processes	E1, E2, E3, E4
		Particles are small balls	E1, E2, E3
		Focus on the concept of "building blocks" and neglect of the concept of "interaction"	E1, E2
	Potential sources of inadequate conceptions	Use of scientific language is a source of misconceptions	E1, E2, E3
		The axes of FDs are misleading	E2, E3
Particle physics can only be treated superficially	Limitations by educational setting	The time used for particle physics could be used otherwise	E1, E3
		Necessary prior knowledge is missing in school-level physics	E1, E3
		It is a challenge for teachers to teach modern physics	E1
	The disciplinary handling of FD is not taught in school	Calculations might be too difficult	E1, E2, E3, E4
		Drawing FDs is challenging	E2
		Some concepts are too difficult for school-level physics	E2

3.1.1. FDs Evoke and Perpetuate Inadequate Conceptions about Particle Physics

The experts discussed several inadequate conceptions, as well as possible sources of these conceptions, that are connected to FDs.

The conception that was mentioned most often as an inadequate conception connected to FDs was that they represent processes embedded in space–time. E1 pointed out that "this form of illustration in the sense of spatiotemporal embedding" should be avoided. However, E3 has "also noticed with teachers" that they "try to read off information from [the] geometry about the movement ... of the particles involved". E2 also clarifies that the interpretation of an FD as a space–time diagram is "only a crutch here" because it cannot "be a real space-time diagram at all", since there is "no concept of path in quantum mechanics". E4 describes it as a "graphic idealization of a process from which one thinks that it actually does not happen that way, but which seems to be the correct technique for bookkeeping".

Close to that idea, the experts said that it is a common misconception that FDs show observable processes. E2 explains that an FD is "not at all an image of a process in that sense" but "a quantum mechanical amplitude" and that "a process is not only described by one FD but by infinitely many". E3 emphasized that "an FD is not a literally readable description of an actually happening process". E1 went on to say that "a physical process, that must be something that I can observe in the world. The rock that falls. And FDs just do not play in this league". He also pointed out that FDs "ostensibly tell stories about what is happening" and expands on this point by emphasizing that quantum field theory, that is the theory underlying elementary particle physics, is "no less strange" than quantum mechanics and that the "form of clarity" suggested by FDs is "not [offered] by our modern theories".

Another inadequate conception closely connected to FDs is the conception of particles as being small balls which physically collide with each other. E2 explained in the context of Compton scattering that even university students still have the conception that "an electron- and a photon-ball meet each other and then there is an elastic collision, and then there is somehow energy- and momentum-transition and then the electron and the photon fly away from each other with a different momentum". E1 compared this to the transport conception of electric current which was "as wrong as only few things" but "still this naïve particle conception could be measured ... so strongly". He also made a point about interaction particles stating that "the idea that a particle mediates something through an exchange" is "pure use of a metaphor" in his eyes. He emphasized that "nothing is exchanged. It is not here before and there afterwards".

Not directly related to FDs, E1 mentioned the challenge that particle physics is often thought of as "searching the fundamental building blocks" while, in fact, "the part-whole relation of modern physics" would be in fact "superposition and mixing". E2 also mentioned this sub-subtheme but was more practical in the sense that he mentioned that he "is not a friend of first introducing all the particles because that only leads to learning them by heart". They shared the concern of conveying a reductionist image of particle physics as the theory of "the fundamental building blocks".

The experts discussed some sources of misconceptions stemming from the FDs. One important source for the origin of these misconceptions is the use of FDs in the everyday life of scientists and hence the use of scientific jargon which is closely related to their use. E1 described this circumstance as "the use of FDs is always embedded in a practice" and "talking about FDs ... is simply jargon". For students, however, "these diagrams ... are simply pure poison" because "there is the danger that school only conveys the jargon but not the practice of course". He stated that "experts talk like topsy-turvy about them" and would use "the most naïve metaphors such as 'comes in', 'comes out', 'get scattered'", but this would all be "embedded into practice which then clears the misconceptions". Besides these general considerations about scientific language, the experts mentioned several specific terms which might be a source of misconceptions, in particular, the term "virtual particle", which E2 calls "very misleading" because "it leads one to say they actually weren't there, as in virtual reality, and that's not the case". A term which was explicitly mentioned as a term used by experts by E2 and E3 but not suitable for students was the term "propagator".

Another source of misconceptions that was mentioned was the drawing of space- and time-axes in FDs. E2 explained that "in the region where the interactions take place, there is no linear order of time anymore ... but only the initial and final state, where the two lines come in and where the two lines go out, there it can be imagined in good approximation as trajectories". He admitted that he was not sure whether "the axes should be omitted in the first place or only the *t*[time]-axis" and suggested to investigate *x*-*t*[space-time]-axes, whether they help or lead to even more misconceptions". E3, on the other hand, suggested that "one should not draw a space axis".

3.1.2. Feynman Diagrams Can Only Be Treated Superficially with High School Students

The second theme in the dimension of challenges concerned the treatment of FDs in in school physics classes. Two subthemes belong to this theme. The first subtheme is related to the framework conditions that might inhibit the appropriate treatment of FDs in the classroom. The second subtheme concerns the lack of practice when handling FDs.

One challenge that was pointed out by E1 was that the discussion of particle physics in school lessons takes time away from the discussion of other curricular topics, since particle physics is often not marked as a mandatory topic in high school physics. On the question of whether FDs or particle physics should be treated in school lessons, he explained that this was also connected with the question of what would be lost as a result. This time "should not be taken away from quantum mechanics, for example". E3 pointed out that "if the connection to theory and to quantum mechanics cannot be made due to lack of time and so forth, then it [the treatment of FDs] shouldn't be done".

E3 also pointed out that for an appropriate introduction of FDs in school lessons, a certain level of "prior knowledge is necessary in order not to be subject to a too literal reading of the diagrams", that is, "prior knowledge of quantum objects in general". E1 also raised the concern that the topic of particle physics might be taught "by teachers who never learned about that topic in their education", since particle physics is often not part of the mandatory courses for teacher training students.

The lack of practice is most prominent in calculations that are too difficult for school. E1 pointed out that particle physics in schools can only convey "an overview knowledge. Even more than in other fields of school physics, students will not learn any calculations, they won't solve any concrete problems". E2 described one experience with high school students when it was too difficult "to explain in one day" to draw a certain conclusion from a calculation. E4 points out that "one cannot hope that one can gain a complete understanding of these theories without diving deep into the mathematics".

E2 also mentioned the experience that "it seems to be more difficult for [university] students to draw FDs than to interpret presented FDs", but he admitted that "this was for reasons [he] doesn't understand".

Moreover, E2 mentioned two concepts which are explicitly too complicated for students, namely the transformation to the center-of-mass inertial system for arguing that certain decays are kinetically forbidden and the mixing via the Cabibbo–Kobayashi–Maskawa matrix.

3.2. Opportunities

Table 2 shows the three themes and their associated subthemes that were attributed to the dimension of 'opportunities'. In contrast to the first dimension, the subthemes are not further divided into sub-subthemes.

Table 2. Themes and subthemes for the dimension opportunities. The order of the subthemes within a theme is according to the frequency of overall mentions.

Theme	Subtheme	Mentioned by
FDs offer a link between particle physics and high school topics	FDs are suited to teach conservation laws	E1, E2, E3
	FDs link particle physics and quantum mechanics	E1, E2, E4
	FDs offer an insight into the use of structurally equivalent representations	E1, E2, E3
	FDs offer an analogy to resonance phenomena in classical oscillations	E2, E3
FDs offer an opportunity to teach different particle physics topics	Outer and inner lines/virtual particles	E1, E2, E3
	Introduction of interaction particles	E1, E2, E3
	Suggestions for educational uses of FDs	E1, E2, E3
	Particle types	E2, E3, E4
	Introduction of pair production and annihilation	E1
FDs offer a connection to current research	FDs help scientists to discuss particle processes	E1, E2, E3, E4
	Particle physics is a showcase for modern science	E1, E2, E3, E4
	Students can find FDs in popular scientific representations	E1, E3

3.2.1. FDs Offer a Link between Particle Physics and Other High School Physics Topics

The experts mentioned in the interviews that FDs are linked in many ways to concepts that occur in school lessons. Most prominently, the concept of charge conservation was judged to be very connectable. E2 described FDs as "a wonderful means to check for conservation laws" and even estimated that "charge conservation ... could definitely be done with younger [than upper secondary] students". Furthermore, E1 uses conservation laws when he is asked to explain a specific FD.

Almost as often as the first subtheme, the experts mentioned that a link between quantum mechanics and particle physics could be established using FDs. E1 emphasized that in the teaching of elementary particle physics, "a certain unity of modern physics" should be presented and that "one should see where one can deepen what one has already learned in quantum mechanics, that this is regarded as [a] unity". E2 pointed out that "one should refer to the other world of quantum physics as early as possible". Moreover, E4 pointed out that "it is important to see that the Standard Model is a quantum theory, that the quantum effects which one has maybe discussed more generally are built into it". More specifically regarding FDs, E1 stated that he would "elucidate the meaning of an FD analogous to the single amplitudes in the double slit, ... that there is one contribution of the wave function, which expresses that the electron goes through one slit". But this would not be a process "which occurs in nature. But the square of the sum of both of these parts corresponds to the observation". E2 summarized this by stating that "FDs don't describe the truth singularly but only the coherent sum describes the truth".

The experts also described the analogy of FDs offering an insight to structural equivalence in analogy to the teaching of electricity. E1 explained that FDs have structural similarities with drawings of electrical circuits. They both have in common that it is important to see "where something is only symbolically noted and how literally a symbolic depiction can be read ... This kind of structural knowledge can be used ... not to fall into the trap to think so vividly". The experts called this notion of invariance concerning different drawings 'topological equivalence'. E3 explained electrical circuits in this sense as "an example where one does not need to focus on the exact geometry in order to analyze what is depicted". E2 expanded on this thought by making the analogy that "the conservation law is the junction rule of currents for example. Current just doesn't get lost. ... And these charge currents are also in the FDs". E1 also made another point connected to the teaching of electricity by stating that particle physics could already be introduced together with electricity "if we want to make the point in particle physics that actually there are no particles, then this could be the conclusion which already informs the teaching of electricity in lower secondary schools".

Another analogy that was highlighted by the experts was between the concept of virtual particles, which are "off mass-shell", and the concept of forced oscillations in mechanics. E2 explained this concept as: "in classical mechanics it is not special at all to have an oscillating system which can be excited with all frequencies and when it is excited with the eigenfrequency I get a high amplitude and it's just like that with virtual particles when I excite them with their eigenfrequency, then the process is very likely because I have a high amplitude". He also connected this concept to the nuclear fusion in the sun, which is happening so slowly because "there all the W-particles ... are highly virtual". This concept might be "fascinating for high school students". E3 also mentioned this analogy and estimated that "it might work well for teachers", while for students it might be more difficult since "forced oscillations ... do not play a role everywhere or only in selected courses".

3.2.2. FDs Offer an Opportunity for Different Particle Physics Topics to Be Taught

While the previous theme was about potential links between particle physics and other high school physics topics, the experts mentioned several aspects of particle physics that could be worthwhile to teach in high school physics. In all of these aspects, FDs play an important role.

One very prominent topic was the distinction between initial, intermediate, and final states. E1 stressed that "the initial state has a meaning, and the final state has a meaning. And when the diagrams do not differ in the initial and final state, then they aren't distinguishable in the measurement". Moreover, then one would be "interested in all possible intermediate processes which can connect them [the initial and final state] according to the rules". This distinction, in his view, has to be "brought home centrally". Further, E2 said that he would "clearly define particles in the initial state and particles in the final state". E3 further explained the distinction between outer lines and inner lines of FDs as for an inner line the "mass-energy-momentum relation ... does not have to be fulfilled" while it has to be fulfilled for an outer line. He also made the connection that "virtual particles are inner lines in FDs". This concept, on the other hand, connects to the analogy of forced oscillations in mechanics mentioned in the previous theme. Connected to the topic of the inner lines is also the topic of "topological equivalence", which was already mentioned in connection with electrical circuits. E3 stressed this topic by explaining that "the temporal order of vertices is not fixed. One can shift vertices and transfer diagrams in other topologically equivalent diagrams, which are then the same diagrams".

Another promising opportunity mentioned is to introduce and further explain the notion of interaction particles (also known as exchange- or messenger-particles or force carriers). E1 even stated that he believes that "the only reason why one should talk about FDs is that one can introduce exchange particles or because exchange particles obtain their metaphorical meaning from this graphical symbolic language". E3 stated the contrasting view that "one needs the concept of exchange- or messenger-particles for that [FDs]" which, in turn, "only make sense ... in this view of fundamental interactions". E2 expands this concept by stating that "when the concept of exchange- or messenger-particles is introduced, then one has to say that these exchange- or messenger-particles couple to charge, that is, they see—in quotation marks—only other objects which carry or own the charge of this interaction".

Two of the experts also mentioned a subtheme on how FDs are connected to different particle types. E2 stated that "it suffices to first only work with the electron, photon, up and down quark, ... electron neutrino, ... W and Z", that is, he only wants to introduce a selection of particles. E3 suggested that "messenger particles could be drawn differently from matter particles to emphasize the distinction". Another important point concerns the antiparticles, which in FDs are depicted with an arrow against the direction of time. E3 acknowledged the inadequate conception of a particle moving backward in time but also stressed that "in his experience, teachers accept [that it's just a symbolic convention]".

One other aspect raised by E1 was the introduction of pair production and particle annihilation, which would be "the main point of relativistic quantum field theory". He explains that "the type of question on which FDs could be an answer to" is about how a detector can detect many particles although only two particles interact with each other.

The experts also made several suggestions on how to introduce FDs in high school classrooms. For example, both E2 and E3 mentioned *Feynman-Rhombino*, a card game which was invented to practice the rules of creating FDs [31]. These rules were described by E2 as "chaining up [basic] vertices to build scattering processes, decay processes, annihilation- and creation processes". E2 also suggested that "in school, one can explain some basic calculation principles", and he described FDs as "a wonderful tool … that I can bring something from the right to the left side, from initial to final state" and by that one can make connections to reaction equations in chemistry or even mathematical equations.

3.2.3. FDs Offer the Opportunity to Discuss Insightful Perspectives on Science

The experts mentioned several aspects where FDs can offer perspectives on science in general and modern physics in particular.

E4 distinguished the roles of FDs "between a pedagogical role to convey the theory and a scientific role to work with the theory as a scientist". The latter was described by E2 from the point of view of a particle physicist. Specifically, for him, FDs are "[a] very clear representation, [a] very helpful, but also very impressively powerful representation". E4 raised the point that "FDs can serve to … also [show] the methodology of particle physics". According to him, physics classes in upper secondary schools should convey "the way of thinking, the methodology which lies behind [the theories and physics]" and "FDs [can be] an extremely good illustration [for this]".

Apart from the connection to the daily work, the experts also mentioned that FDs in particular, and particle physics in general, are showcases for modern science and modern physics. E2 emphasized that "physics has come so far that it has managed to predict what may happen, … what may not happen and even with what probability this or that will happen". This "predictive power" is something that "should definitely be emphasized at school". E4 also described this theory as "the most successful theory that humankind has ever come up with". E3 suggested that students can learn "something about the procedures of and mathematization in theoretical physics" using FDs. E1 suggested using FDs as an opportunity to "[convey] the big message of modern physics … We have theories which work perfectly in a functional sense, but actually we do not know what the objects refer to, at least they are not spatially and temporally embedded". He also raised another point on how particle physics might be useful for students, namely by highlighting "the science-sociological and nature of science aspect that science is a collaborative endeavor".

Another aspect related to the engagement with science is the critical examination of scientific representations, such as in textbooks or popular science texts. E3 explains that "of course, it should also be a learning goal" that "the students can classify popular science literature or sources … " E1 even notes that "the main benefit in the educational context" is that "one can point out widespread misunderstandings" and "send the students on a journey to find misleading or false representations in popular, but also subject-related representations". This could be "a contribution to an experience of autonomy".

4. Discussion

The results from the expert interviews show the potential, but also the difficulties, associated with the introduction of FDs in high school classrooms. The experts had different opinions on whether and how FDs should be used with high school students. Indeed, all experts presented both opportunities and challenges on this matter. Based on the results, four learning objectives for potential learning material with FDs were developed. These are presented and motivated here.

4.1. Learning Objectives

4.1.1. Charge Conservation

The first learning objective is: "Students will be able to use the concept of conservation of charge to determine whether a particle process is possible or not". The subtheme "FDs are a tool to teach about conservation laws" was the most often mentioned opportunity, and all three original experts mentioned it (the fourth expert did not mention this opportunity since the scope of that interview was to clarify discussion points connected to more advanced topics such as the connection to quantum physics). Since conservation laws are arguably one of the most important concepts in physics [45], this learning objective is well suited to also connect FDs to different physics topics such as mechanics or electricity. This connection has also been made by Hoekzema et al. [12], who used a modified version of FDs to teach about symmetries and conservation laws. Additionally, Pascolini and Pietroni [25] used FDs explicitly without their original meaning in the mathematical sense to teach concepts such as conservation laws. Furthermore, there have been different approaches to certain educational activities that help students to use conservation laws with FDs in a playful way [30,31].

We argue that conservation laws are the inevitable entry point to any learning unit dealing with FDs. If the deep connection between conservation laws and FDs is not understood, FDs cannot be used. This first learning objective therefore serves as a prerequisite to all of the following.

4.1.2. Interaction Particles

The second learning objective is: "Students will be able to explain the role of interaction particles and motivate their existence in the Standard Model of particle physics". The experts mentioned interaction particles as one of the most prominent subthemes of the opportunities that FDs can provide for the teaching of particle physics concepts. Lindenau and Kobel [15] argue that the concept of interaction particles is one of the core concepts in the Standard Model of particle physics. A common way to introduce interaction particles is through everyday analogies, such as people on a boat throwing balls and boomerangs at each other, but these tend to promote inadequate conceptions rather than explain physical concepts. We argue that FDs provide a motivation for the concept of interaction particles without going too much into the mathematical details of the Standard Model of particle physics on the one hand and avoiding simple heuristics on the other hand. A more sophisticated way is to connect interactions with fields using the concept of interaction particles [7,28,46]. While this is arguably an adequate way to think about interaction particles—or particles in general—it is also more abstract and thus more difficult to comprehend. However, the concept of interaction particles could be an opportunity to link this idea of how interactions are thought about in particle physics to the concept of the electromagnetic field, which might already be known from earlier school physics.

4.1.3. Superposition

The third learning objective should directly link particle physics to already known quantum physics concepts. It consists of two parts to make it more accessible and relevant to high school education. The first part is: "Students will be able to apply the superposition principle from quantum mechanics to particle processes". Here, students should realize that a process is not only represented by a single diagram, but that several diagrams always contribute to a process. This is in analogy to the double and multiple slit experiment, where several slits contribute equally to the interference pattern. Extending this principle, the second part of this learning objective is: "Students will recognize particle processes as a superposition of infinitely many contributions". Here, students learn that for any FD there can be another FD drawn that represents the same particle process.

Also in the literature, Passon et al. [29] mention the above mentioned analogy to the double slit experiment. Moreover, Allday [7] showcased how this principle is important for learning about the nature of interactions in particle physics. Passon et al. [29] made the connection to the concept of topological equivalence and pointed out structural similarities between FDs and electrical circuits.

This concept is closely connected to a common misunderstanding about FDs, that of representation. As stated above, the debate on whether FDs represent physical processes is almost as old as FDs themselves [21,22]. For certain processes, some diagrams are much more dominant in the calculation; therefore, in these cases, one diagram is a good approximation of what is happening. However, to distinguish between such processes where a single diagram might be a good approximation and processes where several diagrams are needed for a reasonable approximation, a good working knowledge of the corresponding mechanisms, the parameters, and the energies and momenta is needed. Therefore, for high school students, we argue that learning about the general principles might already be challenging enough. For advanced students who already know about the series expansion of a mathematical function, this learning objective could also be linked with an introduction to calculations, but in principle, the learning objective is achievable without addressing this mathematical concept.

4.1.4. Work of Particle Physicists

The experts emphasized that FDs can give an insight into how work is conducted in (theoretical) particle physics. To take advantage of this, the fourth learning objective is: "Students will be able to relate the method of FDs to the work of particle physicists". This learning objective accounts for the fact that a whole theme of opportunities was dedicated to the perspectives on science, and within that, a prominent subtheme was the helpfulness of FDs to scientists. Nevertheless, the helpfulness for scientists alone would not be an argument to use FDs in schools. Some prior knowledge is necessary to use the diagrams in a way scientists use them. We argue that, after achieving the first three learning objectives, students might have acquired enough knowledge to get a glimpse into why and how scientists use FDs. For instance, to achieve this learning objective, one can imagine presenting students with an example of a measurement that does not agree with a theoretical prediction. In this case, the students would have to realize that this could be due, among other things, to the fact that not enough FDs were considered. In this way, the use of FDs can be illustrated in practice.

However, even if all the learning goals were achieved, a full mathematical account would be too ambitious for high school students, as the underlying mathematics requires knowledge of quantum field theory. Nonetheless, as illustrated above, some of the basic underlying calculation principles might be suitable for advanced high school students.

Furthermore, even without going into the details of calculations, a simulation tool could help students understand why the calculation of different diagrams shapes the theoretical predictions for particle physics experiments.

This learning objective is particularly beneficial for learners from a nature of science perspective in the sense that students get to experience first-hand that science is theory-laden by exploring theoretical predictions and getting a glimpse of the tentativeness of science when open questions of modern physics can be examined by FDs (e.g., the high-precision measurement of the magnetic moment anomaly of the muon [47]).

4.2. *Challenges to Address When Teaching with FDs*

While all these learning objectives pose good opportunities for the learning of and with FDs, the inherent challenges that are connected also need to be addressed. These challenges might translate into domain-specific design principles when designing a learning environment that incorporates these learning objectives.

The most prominent theme mentioned within the challenges dimension was that FDs might evoke and perpetuate inadequate conceptions, most notably one of a space–time embedding of particle processes. One discussed source for this inadequate conception was that of the drawing of an FD as a space–time diagram. Among the experts, there was a tendency towards the opinion that while a time axis is useful to denote what is the initial and what is the final state, a space axis does not make any physical sense. Therefore, we argue that FDs should be clearly distinguished from space–time diagrams, or, at least, drawing a space axis should be avoided. Closely connected to this is the conception that particles are small balls that are physically colliding at the interaction points. Both conceptions might also be perpetuated by the depicting particles using lines with arrows. We therefore suggest exploring the possibility to omit the arrows when using FDs in educational contexts. The only function of the arrows in an FD is to distinguish particles from antiparticles. This distinction can be made differently, for example, by using colors or by writing the corresponding symbol. Conceptualizing antiparticles as particles that move backward in time might be a mathematically allowed description when also attributing a negative energy to antiparticles, but it does not represent the physical reality and might evoke misconceptions about the nature of antiparticles.

Another very prominent inadequate conception was that FDs show observable processes. Avoiding this conception is very challenging since part of the popularity of the diagrams is that they seem to "tell a story of what happens" (E1). We argue that pronouncing the third learning objective (particle processes are superpositions of FDs) addresses this inadequate conception.

The most prominent single subtheme in the expert interviews was the challenge that the use of scientific language is a source of inadequate conceptions. This issue might be addressed by a careful use of language in educational contexts [17,48], for example, when speaking of "decay" by clearly explaining that it is in fact a transformation of particles and speaking of "electrical charge" instead of just "charge" to emphasize the fact that there are different charges. However, more research is needed into which conceptions are connected to these different terms and in which way they are beneficial or hindering for learning.

Other challenges touch on the topic of too little time or missing prior knowledge for the meaningful use of FDs in high school physics classes. We propose to address this challenge by linking FDs to concepts that are already known, such as charges, conservation laws, and, for more advanced students, the superposition principle in quantum mechanics or even series expansion in mathematics. If the latter is not yet known by the students, calculations can be omitted in educational settings or substituted by simulation tools instead.

4.3. Outcomes and Limitations

The scope of the study was to explore possible challenges and opportunities of using FDs in high school education. In doing so, this study was the first one to explore the educational use of FDs empirically. Our goal was to interview experts whose opinions are representative of the field. For example, in the interviews we could find the above-mentioned "Feynman-Dyson split" [18,21] to some extent. While E1 stressed the purely mathematical nature of FDs and strongly opposed the notion of a too literal reading of the diagrams, E2 was very open to using the diagrams as a pictorial representation also in a qualitative sense, that is, without immediately referring to the mathematical expression that certain FDs represent. Nevertheless, we have only covered a narrow perspective on the topic, since all four experts came from the German-speaking region (the first three from Germany, the fourth from Switzerland). This might have led to a biased view. However, the small number of experts allowed an in-depth analysis in line with the explorative purpose of the current study. This study in its present form already gives valuable insight for the development of educational material. Moreover, it serves as a basis for future studies with a larger number of experts that can be envisaged to build upon our initial results by conducting a Delphi study [49] and gathering additional perspectives on the educational value of FDs.

4.4. Outlook

The learning objectives and design suggestions developed in this study are used to create learning material for introducing FDs to 14–18-year-olds within a massive open online course (MOOC) on particle physics. The development of this learning material follows the iterative design principles of the design-based research framework [32]. Specifically, design principles based on multimedia learning theories are developed for these learning materials and are combined with more domain-specific design principles resulting from this study. The developed learning material is tested in teaching experiments using eye-tracking data with high school students from the target group. The learning materials will be further developed based on the results of these teaching experiments.

5. Conclusions

The presented study systematically analyzed the risk and potential that lies in teaching about particle physics to high school students using FDs. We have found two general challenges and three major opportunities that are connected to this peculiar form of representation within particle physics. We argue that the opportunities we have found are calls for using FDs in teaching particle physics, even beyond their original function as mere calculation tools. Indeed, FDs can serve as a tool to connect particle physics to already known concepts, such as charges and conservation laws on the one hand and to other advanced topics such as quantum mechanics on the other hand. Since FDs are used by particle physicists every day, their educational use can offer high school students a window into the daily work of particle physicists. The challenges we have found can be addressed by domain-specific design principles for developing a learning environment for and with FDs.

We are confident that it is useful to introduce high school students to FDs, not least because FDs are part of popular representations of particle physics. As one of the experts who took part in this study stated, one can "send students on a journey to find misleading or wrong depictions in popular or even specialized media … that can be a contribution to an experience of autonomy". This could be a major strength of using this form of representation in physics classrooms.

Supplementary Materials: The following supporting information can be downloaded at: https://www.mdpi.com/article/10.3390/physics4040085/s1, Table S1: Coding Manual; Document S2: Interview guideline E1, E2, E3; Document S3: Interview guideline E4.

Author Contributions: Conceptualization, M.N.D., J.W., P.K. and A.M.; methodology, M.N.D. and J.W.; validation, J.W.; investigation, M.N.D.; writing—original draft preparation, M.N.D.; writing—review and editing, M.N.D., J.W., P.K. and A.M.; supervision, P.K., A.M., S.M.S. and J.W.; project administration, S.M.S. All authors have read and agreed to the published version of the manuscript.

Funding: This work has been sponsored by the Wolfgang Gentner Programme of the German Federal Ministry of Education and Research (grant no. 13E18CHA).

Institutional Review Board Statement: Ethical review and approval were waived for this study, due to the fact that the study was in accordance with the Requirements of our institution: https://cds.cern.ch/record/2651311/files/Circ_Op_Angl_No11_Rev0.pdf (accessed on 25 October 2022).

Informed Consent Statement: The subjects gave their informed consent that their data will be used for research purposes.

Data Availability Statement: The data presented in this study, namely the interview segments and, if justified, the full interview transcripts, are available on request from the corresponding author. The data are not publicly available due to the personality rights of the experts.

Acknowledgments: Thank you to the four experts who contributed to this study. Many thanks to Larissa Hahn and Simon Lahme (both University of Göttingen) for a thorough discussion of the coding scheme and for serving as interraters. Moreover, thanks to Marlene Breither (University of Göttingen) for serving as interrater.

Conflicts of Interest: The authors declare no conflict of interest. The funders had no role in the design of the study; in the collection, analyses, or interpretation of data; in the writing of the manuscript; or in the decision to publish the results.

References

1. Swinbank, E. Particle Physics: A new course for schools and colleges. *Phys. Educ.* **1992**, *27*, 87–91. [CrossRef]
2. Aubrecht, G.J. Report on the conference on the teaching of modern physics. *Phys. Teach.* **1986**, *24*, 540–547. [CrossRef]
3. Sjøberg, S.; Schreiner, C. Results and perspectives from the Rose project. In *Science Education Research and Practice in Europe: Retrospective and Prospective*; Jorde, D., Dillon, J., Eds.; Sense Publishers: Rotterdam, The Netherlands, 2012; pp. 203–236. [CrossRef]
4. Barlow, R. Particle Physics: From School to University. *Phys. Educ.* **1992**, *27*, 92–95. [CrossRef]
5. Kobel, M. High school students' exposure to modern particle physics. *Europhys. News* **2003**, *34*, 108–110. [CrossRef]
6. Tuzón, P.; Solbes, J. Particle physics in high school: A diagnose study. *PLoS ONE* **2016**, *11*, e0156526. [CrossRef]
7. Allday, J. The nature of force in particle physics. *Phys. Educ.* **1997**, *32*, 327. [CrossRef]
8. Zügge, T.; Passon, O. Das Wuppertaler Curriculum der Elementarteilchenphysik. In *Kohärenz im Unterricht der Elementarteilchenphysik*; Passon, O., Zügge, T., Grebe-Ellis, J., Eds.; Springer Spektrum: Berlin/Heidelberg, Germany, 2020; pp. 121–141. [CrossRef]
9. Kobel, M.; Bilow, U.; Lindenau, P.; Schorn, B. *Teilchenphysik: Ladungen, Wechselwirkungen Und Teilchen, 3. Auflage.*; Joachim Herz Stiftung: Hamburg, Germany, 2018. [CrossRef]
10. Polen, C. Particle physics: An essential and engaging part of the high school physics program. *Phys. Teach.* **2019**, *57*, 320–322. [CrossRef]
11. Van de Berg, E.; Hoekzema, D. Teaching conservation laws, symmetries and elementary particles with fast feedback. *Phys. Educ.* **2006**, *41*, 47–56. [CrossRef]
12. Hoekzema, D.; Schooten, G.; van den Berg, E.; Lijnse, P. Conservation laws, symmetries, and elementary particles. *Phys. Teach.* **2005**, *43*, 266–271. [CrossRef]
13. Michelini, M. Innovation of curriculum and frontiers of fundamental physics in secondary school: Research-based proposals. In *Fundamental Physics and Physics Education Research*; Sidharth, B.G., Murillo, J.C., Michelini, M., Perea, C., Eds.; Springer International Publishing: Cham, Switzerland, 2021; pp. 101–116. [CrossRef]
14. Passon, O. Mystifizierung der Quantenmechanik und Trivialisierung der Teilchenphysik. In *Kohärenz im Unterricht der Elementarteilchenphysik: Tagungsband des Symposiums zur Didaktik der Teilchenphysik, Wuppertal 2018*; Passon, O., Zügge, T., Grebe-Ellis, J., Eds.; Springer Spectrum: Berlin/Heidelberg, Germany, 2020; pp. 79–90. [CrossRef]
15. Lindenau, P.; Kobel, M. Introducing an innovative approach of teaching the Standard Model of particle physics at high school. *J. Phys. Conf. Ser.* **2019**, *1287*, 012045. [CrossRef]
16. Gourlay, H. Learning about A level physics students' understandings of particle physics using concept mapping. *Phys. Educ.* **2016**, *52*, 014001. [CrossRef]
17. Wiener, G.J.; Schmeling, S.M.; Hopf, M. Can grade-6 students understand quarks? Probing Acceptance of the subatomic structure of matter with 12-year-olds. *Eur. J. Sci. Math. Educ.* **2015**, *3*, 313–322. [CrossRef]
18. Kaiser, D. Drawing Theories Apart: The Dispersion of Feynman Diagrams in Postwar Physics; University of Chicago Press. 2005. Available online: https://academic.oup.com/chicago-scholarship-online/book/23068 (accessed on 10 October 2022).
19. Dyson, F.J. The radiation theories of Tomonaga, Schwinger, and Feynman. *Phys. Rev.* **1949**, *75*, 486–502. [CrossRef]
20. Feynman, R.P. Space-time approach to quantum electrodynamics. *Phys. Rev.* **1949**, *76*, 769–789. [CrossRef]
21. Passon, O. On the interpretation of Feynman diagrams, or, did the LHC experiments observe $H \rightarrow \Gamma\gamma$? *Eur. J. Philos. Sci.* **2019**, *9*, 20. [CrossRef]
22. Meynell, L. Why Feynman diagrams represent. *Int. Stud. Philos. Sci.* **2008**, *22*, 39–59. [CrossRef]
23. Dorato, M.; Rossanese, E. The nature of representation in Feynman diagrams. *Perspect. Sci.* **2018**, *26*, 443–458. [CrossRef]
24. Passon, O.; Zügge, T.; Grebe-Ellis, J. Pitfalls in the teaching of elementary particle physics. *Phys. Educ.* **2018**, *54*, 015014. [CrossRef]
25. Pascolini, A.; Pietroni, M. Feynman diagrams as metaphors: Borrowing the particle physicist's imagery for science communication purposes. *Phys. Educ.* **2002**, *37*, 324. [CrossRef]
26. Jones, G.T. The Uncertainty Principle, virtual particles and real forces. *Phys. Educ.* **2002**, *37*, 223–233. [CrossRef]
27. Woithe, J.; Wiener, G.J.; der Veken, F.F.V. Let's have a coffee with the Standard Model of particle physics! *Phys. Educ.* **2017**, *52*, 034001. [CrossRef]
28. Lambourne, R. Predicting the physics of particles. *Phys. Educ.* **1992**, *27*, 71. [CrossRef]
29. Passon, O.; Lindenau, P.; Kobel, M. Von Feynman-Diagrammen und Stromkreisen. *Unterr. Phys.* **2020**, *180*. Available online: https://www.friedrich-verlag.de/physik/quantenphysik/von-feynman-diagrammen-und-stromkreisen-7480 (accessed on 10 October 2022).
30. Harlander, R.V.; Klein, S.Y.; Lipp, M. FeynGame. *Comput. Phys. Commun.* **2020**, *256*, 107465. [CrossRef]
31. Lindenau, P.; Winkler, O. Feynman-Rhombino. *Unterr. Phys.* **2020**, *180*. Available online: https://www.friedrich-verlag.de/physik/quantenphysik/feynman-rhombino-7486 (accessed on 10 October 2022).
32. Baumgartner, E. et al. [The Design-Based Research Collective] Design-based research: An emerging paradigm for educational inquiry. *Educ. Res.* **2003**, *32*, 5–8. [CrossRef]
33. Duit, R.; Komorek, M. Understanding the basic ideas of chaos-theory in a study of limited predictability. *Int. J. Sci. Educ.* **1997**, *19*, 247–264. [CrossRef]

34. Kersting, M.; Henriksen, E.K.; Bøe, M.V.; Angell, C. General relativity in upper secondary school: Design and evaluation of an online learning environment using the model of educational reconstruction. *Phys. Rev. Phys. Educ. Res.* **2018**, *14*, 010130. [CrossRef]
35. Kamphorst, F.; Vollebregt, M.J.; Savelsbergh, E.R.; van Joolingen, W.R. An educational reconstruction of special relativity theory for secondary education. *Sci. Educ.* **2021**. [CrossRef]
36. Anderson, T.; Shattuck, J. Design-based research: A decade of progress in education research? *Educ. Res.* **2012**, *41*, 16–25. [CrossRef]
37. Haagen-Schützenhöfer, C.; Hopf, M. Design-based research as a model for systematic curriculum development: The example of a curriculum for introductory optics. *Phys. Rev. Phys. Educ. Res.* **2020**, *16*, 020152. [CrossRef]
38. Barab, S.; Squire, K. Design-based research: Putting a stake in the ground. *J. Learn. Sci.* **2004**, *31*, 1–14. [CrossRef]
39. Duit, R.; Gropengießer, H.; Kattmann, U.; Komorek, M.; Parchmann, I. The model of educational reconstruction—A framework for improving teaching and learning science. In *Science Education Research and Practice in Europe: Retrospective and Prospective*; Jorde, D., Dillon, J., Eds.; SensePublishers: Rotterdam, The Netherlands, 2012; pp. 13–37. [CrossRef]
40. Bogner, A.; Littig, B.; Menz, W. *Interviews mit Experten: Eine praxisorientierte Einführung*; Springer: Wiesbaden, Germany, 2014. [CrossRef]
41. Wiener, G.J.; Schmeling, S.M.; Hopf, M. The technique of probing acceptance as a tool for teachers' professional development: A PCK study. *J. Res. Sci. Teach.* **2018**, *55*, 849–875. [CrossRef]
42. Burde, J.-P.; Wilhelm, T. The electron gas model as an introduction to electricity in middle school science. In *Electronic Proceedings of the ESERA 2015 Conference: Science education research: Engaging Learners for a Sustainable Future*; Lavonen, J., Juuti, K., Lampiselkä, J., Uitto, A., Hahl, K., Eds.; *Part 1: Conceptual Understanding*; Finalayson, O., Pinto, R., Eds.; University of Helsinki: Helsinki, Finland, 2016; pp. 26–36. Available online: https://www.dropbox.com/s/wrb22pgeociw8dr/eBook2015_Part_1_links.pdf (accessed on 10 October 2022).
43. Braun, V.; Clarke, V. Using thematic analysis in psychology. *Qual. Res. Psychol.* **2006**, *3*, 77–101. [CrossRef]
44. Hayes, A.F.; Krippendorff, K. Answering the call for a standard reliability measure for coding data. *Commun. Meth. Meas.* **2007**, *1*, 77–89. [CrossRef]
45. Hill, C.T.; Lederman, L.M. Teaching symmetry in the introductory physics curriculum. *Phys. Teach.* **2000**, *38*, 348–353. [CrossRef]
46. Daniel, M. Particles, Feynman diagrams and all that. *Phys. Educ.* **2006**, *41*, 119–129. [CrossRef]
47. Wolchover, N. 'Last Hope' Experiment Finds Evidence for Unknown Particles. Quanta Magazine. 7 April 2021. Available online: https://www.quantamagazine.org/last-hope-experiment-finds-evidence-for-unknown-particles-20210407/ (accessed on 10 October 2022).
48. Wiener, G.J.; Schmeling, S.M.; Hopf, M. Introducing 12 year-olds to elementary particles. *Phys. Educ.* **2017**, *52*, 044001. [CrossRef]
49. Osborne, J.; Collins, S.; Ratcliffe, M.; Millar, R.; Duschl, R. What "ideas-about-science" should be taught in school science? A Delphi study of the expert community. *J. Res. Sci. Teach.* **2003**, *40*, 692–720. [CrossRef]

MDPI AG
Grosspeteranlage 5
4052 Basel
Switzerland
Tel.: +41 61 683 77 34

Physics Editorial Office
E-mail: physics@mdpi.com
www.mdpi.com/journal/physics

www.ingramcontent.com/pod-product-compliance
Lightning Source LLC
Chambersburg PA
CBHW061602210926
51850CB00009B/2192
9783725856794